Environmental Sociology Now

Environmental Sociology Now

EDITED BY JORDAN FOX, IAN CARRILLO,
J. P. SAPINSKI, AND DIANA STUART

University of California Press

University of California Press
Oakland, California

Library of Congress Cataloging-in-Publication Data

Names: Fox, Jordan editor | Carrillo, Ian editor | Sapinski, J. P. editor | Stuart, Diana (Diana Lynne), 1979- editor
Title: Environmental sociology now / edited by Jordan Fox, Ian Carrillo, J. P. Sapinski, and Diana Stuart.
Description: Oakland, California : University of California Press, [2026] | Includes bibliographical references and index.
Identifiers: LCCN 2025042206 (print) | LCCN 2025042207 (ebook) | ISBN 9780520421240 cloth | ISBN 9780520421257 paperback | ISBN 9780520421264 epub
Subjects: LCSH: Environmental sociology
Classification: LCC GE195 .E589 2026 (print) | LCC GE195 (ebook)
LC record available at https://lccn.loc.gov/2025042206
LC ebook record available at https://lccn.loc.gov/2025042207

GPSR Authorized Representative: Easy Access System Europe, Mustamäe tee 50, 10621 Tallinn, Estonia, gpsr.requests@easproject.com

35 34 33 32 31 30 29 28 27 26
10 9 8 7 6 5 4 3 2 1

Contents

Illustrations

Acknowledgments

As the title of this volume suggests, its original inspiration was Claudio E. Benzecry, Monika Krause, and Isaac Ariail Reed's tremendous 2017 volume *Social Theory Now* (University of Chicago Press). Isaac Reed, in particular, provided useful insight during *Environmental Sociology Now*'s initial, COVID-hazed planning stages. The editors would also like to thank all the contributors to this volume for their time, openness, patience, and acumen. The University of California Press has moreover been a dream to work with. We would like to personally thank Stacy Eisenstark, Sam Warren, Emily Park, and Gabriel Bartlett, as well as our insightful reviewers, for their respective contributions at the beginning and the end of the publishing process. Our greatest UC Press gratitude, however, is reserved for Naja Pulliam Collins, whose tenacious belief in the project and attention to detail provided a sense of security in what was a new sort of venture for each of us. Thanks also to Richard York, Caleb Scoville, Jack Zinda, Kari Norgaard, Tad Skotnicki, Arthur Russel Wallace, Jim Elliott, Erin Hatton, Brett Clark, Anupriya Pandey, Lauren Richter, Matt Norton, Jaume Franquesa, Hannah Holleman, Surahbi Pant, Julian Montague, Phil Campanile, and Scott Frickel for their various support throughout.

Above all, Jordan Fox dedicates his contribution to Lillian Florence Wischerath, Henry Fox Wischerath, and Melissa Dorothy Wischerath; Ian dedicates his contribution to Annabel and Torino; J.P.'s contribution is dedicated to Alder and Rose; Diana would like to dedicate her contribution to Iris.

Introduction

Jordan Fox, Ian Carrillo, J.P. Sapinski, and Diana Stuart

SETTING OUT

Environmental sociology is currently in a critical phase of transformation. While its decades-long history has never been one of stagnation, it is now, more than ever, driven by change. This book illustrates many of the areas of transformation that now characterize the subdiscipline in North America. It contains original essays from emerging scholars who are building new ideas and laying new foundations. It is ambitious in its attempt to capture this dynamism, a goal that by nature will be incomplete, imprecise, and ongoing.

The driving forces of the current transformations in North American environmental sociology stem from multiple sources. For example, while environmental sociology has always, in some sense, been interdisciplinary, disciplinary boundaries are today less relevant than ever. Contemporary environmental sociologists are drawing from other social science disciplines as well as from the natural sciences in new ways, unraveling boundaries in order to foster collaboration, insights, and broader understandings. An increasing number of scholars are publishing in interdisciplinary journals, and many of the lines between environmental sociology, environmental anthropology, geography, environmental studies, and other traditions with similar concerns are now blurrier than before.

In addition, environmental sociologists are today pushing to include ever more of the nonhuman world. At its core, **environmental sociology** has always been about examining the relationships between humans and the entities and processes often referred to as "nature" or "the environment." While, at first blush, these concepts seem easy to understand, it is actually deceivingly difficult to pin down exactly what constitutes "nature," "the environment," or even "society," especially when we try to examine the lines between these concepts (See chapter 1 of this volume). This difficulty raises some important questions, some of which have been a part of environmental sociology for some time. For instance: Just how are we supposed to analyze social processes and environmental processes at the same time, exactly? Or, is it even possible to study societies as if they are not related to the ecosystems, climates, and other "natural" processes through which societies change? Some questions, though, are relatively new. For example: How are social processes like neocolonialism and systemic racism currently changing the ways we define what counts as an environmental problem? Or, how might recent insights into animal cognition force us to revise the ways we think about human-animal relationships? To answer these and other questions, old and new, environmental sociologists now, more than ever, home in on examining the associations between humans and other entities, beings, processes, and structures that underpin the existence of societies. For example, Gould and Lewis (2020) describe environmental sociology as "the study of how social systems interact with ecosystems" (ix). Lockie (2015) explains that environmental sociology is "the application of our sociological imaginations to the connections among people, institutions, technologies and ecosystems that make society possible" (140). Notice how each definition includes ecosystems, referring to the interconnectedness of all living and nonliving entities. In the age of climate change and other global ecological and existential crises, it is clear that we should not only seek to understand these ecological and other relationships but continue to question their boundaries and complexion.

As our questions shift, so do our bases for answering them. Environmental sociologists are now using a wider diversity of methods in conversation with a wider diversity of theories to examine the boundaries and complexion of socio-ecological relationships. While, again, environmental sociology has never been tied to any one theoretical or methodological perspective, the influence of other disciplines, combined with the diversification of method and theory in sociology itself, has led to an expansion of theoretical backgrounds and methodological approaches in environmental sociology. As the chapters in this volume attest, environmental sociologists continue to engage theoretical developments not only across sociology but beyond it—for instance, in the philosophy of science (chapter 1), queer theory (chapter 3), and Indigenous studies (chapter 6), just as they are inspired by a wide diversity of novel methodological techniques, from the quantitative (chapter 9), to the participatory (chapter 4), to the agroecological (chapter 12).

Perhaps the most important change in the diversity of the subdiscipline, though, is the diversification of practicing environmental sociologists.

While the subdiscipline remains overwhelmingly white (Mascarenhas et al. 2016; Liévanos 2021), environmental sociologists today come from a greater diversity of racial, gendered, economic, political, and institutional backgrounds than before in the United States and Canada. Indeed, the essays in this volume are designed to reflect the multiple and interacting senses and manners in which North American environmental sociology has become more diverse.

Lastly, social and environmental crises have escalated to levels that not only make the work of environmental sociologists more critical but also reshape the scope and purpose of present and future research. These crises are multiple and intersecting, spurred by climate change, biodiversity loss, economic and social inequality, environmental migration, inadequate healthcare, and more. Moreover, it has become crystal clear that these problems are not separate from each other, as some people, much more than others, face a higher burden of risks related to multiple threats. For example, low-income communities of color living near sources of pollution and exposed to poor air quality are more likely to also reside in areas prone to fire, storms, or floods from climate-related disasters. These communities may, in addition, face racially based violence and inadequate access to healthcare and healthy food. Hence, environmental justice has become central to the field, and environmental sociology is embracing intersectionality as it enters into dialogue with the issues brought to the fore by social justice movements such as Black Lives Matter and Indigenous resurgence. Environmental sociologists continue to conduct research that reveals intersecting threats, exposes injustices, and improves environmental and social outcomes, especially for those facing the greatest burdens of harm.

This book presents a wide range of essays from early and mid-career environmental sociologists based in the United States and Canada who directly embrace these contemporary changes in environmental sociology and our environmental relationships more broadly. Even though nearly every author comes from a different perspective or set of interests, each essay is nonetheless centered around the question, "What does a more interdisciplinary, more diverse, and more justice-oriented environmental sociology look like, and what does that mean for our collective future?" The answers provided in the chapters that follow both build on and diverge from past work. Its authors often combine environmental sociology's past with new pathways for research, scholarship, and activism. As we do indeed face increasing and global-level existential threats, environmental sociologists must remain open to whatever can best help us analyze and prepare for the challenges ahead.

THE PAST, PRESENT, AND FUTURE OF ENVIRONMENTAL SOCIOLOGY

Past: Changes Inside, Changes Outside

This is not a book that chronicles the history of North American environmental sociology. Nevertheless, it is helpful to begin by noting two related changes that took place in the 1970s—one change within the discipline of sociology itself and one change beyond the discipline—that not only created the space for a formal environmental sociology to emerge but continue to influence the direction of environmental sociology today.

The first, internal, change derived from fundamental questions that were being raised regarding how mainstream sociology—or sociological work produced from core sociology departments, published in the largest and most prestigious journals, and with the broadest reach outside the discipline—was conducted. These questions were raised by a host of scholars from outside that mainstream, including feminists, Black radical scholars, Marxists, poststructural theorists, and others, who, despite profound disagreements on other topics, largely agreed that mainstream sociology's basic strategies for thinking about and investigating social reality were inadequate. The problem with these strategies was that they assumed social reality to be structured through fundamental sets of social laws, sometimes called "covering laws," that form and interact in largely the same manner in most times and places. And, according to these outside scholars, this simply cannot be the case. Social reality is too complex, too overdetermined, too dependent upon its temporal and spatial context, and too weighed down by the legacies of patriarchy, racism, capitalism, and colonialism to be reduced to such fundamental, ahistorical, a-spatial, knowable, systems. For this reason, sociology should quit trying to deduce any social laws and be far more concerned with investigating how social reality, as well as our investigations of it, are power-laden, contingent, and relational. According to these forces outside mainstream sociology, in other words, sociologists needed to rethink the fundamentals of what sociology is and how it was conducted. While it is arguable whether or not this challenge was successful, it is inarguable that it created more space in sociology for scholars to think differently than before (for an overview, see Steinmetz 2005). One of the subdisciplines that rose to fill this space was environmental sociology.

The second change, one that occurred well beyond the discipline, was the development of the modern environmental movement in North America and Europe. Concern about **environmental degradation**, or the social act of rendering a specific set of ecosystems or broader biophysical systems dramatically less amenable to life, was certainly not new, nor was it unique to the western world. Yet in the 1960s and 1970s it became a focal point of many Western activist movements like it had not before in recent history. And some sociologists, many of whom were environmental activists in addition to being sociologists, concluded that building a broader sociological community explicitly focused on the ways that contemporary social processes were driving environmental degradation was both ethically and analytically necessary.

These two changes—the opening up of the discipline to new ways of thinking and the rise of the environmental movement in the Western world—set the stage for what became environmental sociology. The first explicitly environmental sociological arguments were indeed centered around these changes. For instance, early adherents argued that the entire discipline of sociology should be completely rebuilt in order to appreciate the nonhuman actors and processes that were traditionally excluded from mainstream sociology (e.g., Catton and Dunlap 1978). They focused on examining the ways in which humans are not beyond the influence of environmental problems, processes, or constraints but are instead indelibly interactive with environments in complex ways that we still cannot fully appreciate. Moreover, many implored sociologists to join the ranks of the leading environmental activists of the time in order to harness the power of a distinctly sociological analysis of environmental issues.

Today, environmental sociology looks different than it did in those early days. There is now an almost universal awareness of what could be called the proto-environmental sociology impulses of early sociologists and related figures such as W.E.B. Du Bois, Karl Marx, and Max Weber, a more nuanced appreciation of how environmental sociologists should and should not include natural science in their work, a greater emphasis on justice, a far more profound embrace of interdisciplinarity, and much else (Holleman 2021; Pellow 2017; Porcelli and Besek 2021; Stuart 2016). Meanwhile, at present it is not unusual for mainstream sociologists to place environmental problems front and center, and some even cautiously welcome the work of environmental sociologists. Nevertheless, environmental sociology is still in its early days, with many more changes to come.

Present: What Holds Environmental Sociology Together?

Few areas of social science are as wide-ranging as environmental sociology. For what environmental sociologists have done from the start is demand that an already broad field of inquiry—**sociology** or the study of social processes—also include environmental processes and thus become even broader. Considering the sheer number of things and processes that fall under its umbrella, what, then, holds environmental sociology together?

To answer this question, it is useful to start with what is *not* holding environmental sociology together. It is not held together by any one concept. To say it is concerned with concepts such as "society," "nature," and/or "environment," or all three, would be accurate (see, for example, chapter 1 in this volume) but overly imprecise. What these concepts even mean is up for vigorous debate. Most environmental sociologists indeed problematize both these concepts as well as material and analytical distinctions between them, rendering them too complicated and too encompassing to serve as any kind of legitimate foundation (e.g., Salleh 2017).

Nor is environmental sociology held together by any one problem. Environmental sociologists are concerned with pollution (Grant et al. 2020), biodiversity loss (Besek and York 2019), the state and environmental protection (Rea and Frickel 2023), forest depletion (Rudel 2009), the corporatization of agriculture (Leguizamón 2019), scientific classification and the law (Scoville 2022), green gentrification (Gould and Lewis 2016), and, perhaps most of all, climate change (Dunlap and Brulle 2015). Moreover, these and other problems are interactive: they work through each other.

Nor is environmental sociology held together by any methodological or theoretical perspective. Environmental sociologists use quantitative methods (Greiner 2022), historical methods (Baker 2024), interview-based methods (Ergas 2021), ethnographic methods (Luna 2017), mixed-methods (Dryden and McCumber 2017), or some combination of these. And they hold a wide variation of theoretical commitments. Not only are there active feminist (Ducre 2018), decolonial (Gutiérrez-Zamora, Mustalahti, and García-Osorio 2022), Marxist (Auerbach and Clark 2018), actor-network (Rath and Swain 2024), world polity (Theis 2020), and other theoretical traditions; many have built novel theoretical perspectives from these geared explicitly toward environmental sociological literature (see chapter 2 in this volume).

In turn, what *does* hold environmental sociology together is the same tension that has always held it together. On one side, environmental sociologists are committed to the idea that the study of our environmental relationships cannot be left to environmental scientists alone. For instance, problems like climate change and biodiversity loss are not simply environmental problems for environmental scientists to study but also social problems that require social science analysis. These environmental problems are co-produced through networked and power-laden social processes—for example, the social act of burning fossil fuel for capitalist production and worker mobility drives climate change and the social act of clearing a rainforest for growing feed grains helps to drive biodiversity loss. And these environmental problems also affect different people differently, rendering any understanding of them incomplete without consideration of social inequality. Finally, the ways environmental scientists understand what even constitutes what is or is not "environmental" is mediated through their own institutional, political, and other forms of social histories. These social histories should also be in our minds whenever we perform our investigations. In all, to figure out what dynamics are driving our environmental relationships necessitates sociological analysis. Environmental science, by itself, is not enough.

If our environmental relationships are not simply environmental, the other side of this tension is a commitment to how our social relationships are not simply social. Just as a gender scholar would (rightly) assert that gender is foundational to our understanding of most contemporary social processes, or a racial studies scholar would (again, rightly) say that race is foundational to our understanding of most contemporary social processes, environmental sociologists focus on just how much of our social reality is coproduced through wide constellations of environmental processes. While environmental processes never determine social processes, our social worlds nonetheless cannot be fully known absent consideration of environmental relationships. From the materials that make up the book or electronic device on which you are reading these words, to the clothes you are wearing, to your most recent meal, to the contemporary inequalities in the place you presently live, to the ways these things and more both affect and are affected by changing climates, the lives and deaths of other species, ecosystem history, and much else—social processes are fundamentally also environmental processes.

What holds environmental sociology together is that, on the one hand, our environmental relationships require sociological analysis and, on the other, social processes are also environmental processes.

Future: Changes Inside, Changes Outside (Reprise)

Beyond this shared tension, environmental sociology is becoming ever more varied. This is not necessarily a bad thing. It is simply reflective of an emergent conversation as well as an engagement with an ever-changing world.

As before, environmental sociology is transforming through changes that are both internal to sociology as well as outside sociology.

Internally, environmental sociology is today at a crossroads. On one side, its influence is spreading throughout sociology and beyond, and colleges and universities are hiring ever more environmental sociologists to keep up with demand for both graduate and undergraduate classes. On the other side, new voices are building on past work to transform what counts as environmental sociology. Emerging environmental sociologists are focusing explicitly on intellectual and social diversity (e.g., Murphy 2020), drawing from and even joining conversations in related disciplines (e.g., Porcelli and Besek 2022), refocusing the conversation on issues of justice (e.g., Liévanos et al. 2021), and more. As a result, different questions—for instance, "what does a postcolonial environmental perspective look like?" and "what is the difference between political ecology and environmental sociology, if any?"—are driving inquiry forward. In other words, as environmental sociology is growing in relevance, a fresh generation of scholars are shifting the scholarly terrain beneath it.

Many external influences are also changing the course of environmental sociological scholarship, reinforcing the above internal changes. Two are particularly relevant. The first is the changing nature of progressive activism. Progressive activists are today more influenced by racial justice, democratic socialism, Indigenous resistance and resurgence, and environmental justice than even ten years ago, and these concerns are reflected within environmental sociological debate. Scholars are more concerned with making active connections to a wider variety of social movements than before. Second, it is clear that environmental processes will become not only more dynamic in coming decades, as changes in climate and land use already fluctuate more than before, but also interactive. The work of untangling how these processes have unfolded in the past, as well as how they are likely to unfold in the future, has demanded that environmental sociologists broaden their toolkits, whether it be making new alliances or rethinking the boundaries of current conversations, to better explain these contemporary changes.

If environmental sociology was born through a mixture of questions, namely, how to open up sociology to better incorporate environmental processes and how to engage environmental and related forms of activism, a similar mixture will likely drive environmental sociology's future.

BUILDING *ENVIRONMENTAL SOCIOLOGY NOW*

This volume seeks to open up a dialogue by bringing together North American scholars whose work is grounded in multiple strands of environmental sociology. It does not pretend to reveal any comprehensive mutual understandings but rather seeks to function as a space where ideas can interact. In approaching contributing authors, we encouraged them to freely explore arguments and futures in their respective fields of expertise. As such, as you read across this book's chapters, you might notice that some authors disagree on some rather important points. We see this variation as an asset, one that enriches debates and challenges us to critically consider what environmental sociology looks like more than two decades into the twenty-first century. Moreover, this diversity of perspectives also helps us to be cognizant of the tendency for papers to claim novelty when that novelty exists only within a particular tradition (while scholars in other traditions have long engaged those same issues). Of course, work by scholars outside the Western world also furthers a diversity of perspectives. While we do acknowledge that this volume is very much US-centric, we strongly encourage readers to widen their outlook to environmental sociologies beyond the confines of the continent.[1]

We furthermore understand that all knowledge is situated, including the choices made by us, the editors, when bringing this book together. Not only do the essays herein center on a North American context, they also more broadly stem from diverse, specific, Western positionalities. While one should always be careful not to essentialize what are very complex, contextual, and relational personal histories, the positionalities of the editors and the chapter authors—which include Latine, minority francophone, female, settler, white, Black, Reform Jewish, various class backgrounds, and many intersectional others—impact the intellectual labor embodied in the structure of the book and the diverse themes of each chapter.

As we assembled this volume's contributing chapters and glossary (note that bolded words in text can be found in the glossary), we sought insights from a diversity of junior and mid-career scholars with expertise in a specific area of environmental sociology. We asked them to reflect on their expertise, as well as their creativity, in order to answer the following prompt: "What does a more interdisciplinary, more diverse, and more justice-oriented environmental sociology look like, and what does that mean for our collective future?" The following paragraphs briefly describe why we asked chapter authors to focus on "interdisciplinarity," "diversity," and "justice," as well as why we asked them to embrace "creativity."

INTERDISCIPLINARITY

Sociology originated as the study of how social actors and relationships influence social outcomes, and for some time there remained a resistance to including nonsocial or nonhuman entities and processes in sociological studies. Yet, in environmental sociology, engaging with the nonhuman world in research and scholarship was a necessity (Stuart 2016). This work has evolved to include new ways of understanding human and nonhuman relationships, as environmental sociologists have continued to develop new frameworks for analysis while also engaging with scholarship from other disciplines in both the social and natural sciences—both exporting and importing frameworks and ideas across disciplinary boundaries (Lidskog and Waterton 2016). Rather than staying within disciplinary silos, more environmental sociologists are calling for a loosening of disciplinary boundaries to increase our capacity to see and understand environmental problems and to work together to identify solutions (Kasper 2016).

Today, environmental sociologists are increasingly embracing work from diverse disciplines. We see this engagement with work outside sociology as a net positive, as there are plenty of topical and theoretical overlaps and much to be gained through reading other work and collaborating across disciplines. For example, political ecology—a subdiscipline in geography—has much in common with environmental sociology, especially in recent years. In both cases, scholars are focusing on power relations, marginalization, social justice, and drivers of environmental degradation from local to global scales. While we would not argue for political ecology and environmental sociology to become one—they have distinct histories that merit mutual respect—we do believe that cross-fertilizations with political ecology, as well as with other disciplines from conservation biology to human ecology, continue to be a good thing for environmental sociology on the whole.

These efforts foster new collaborations and broaden discussions that aid our mutual efforts to understand and address the social drivers of the climate and biodiversity crises as well as other issues. These conversations and collaborations are more commonly encouraged and will continue to be critical for addressing escalating challenges. For many environmental sociologists, especially those practicing "public sociology," making significant contributions toward positive social change now matters much more than staying within disciplinary boundaries (Caniglia et al. 2021).

DIVERSITY

A recent study by the American Sociological Association's Section on Environmental Sociology, the principal governing body for environmental sociology in the United States, highlights a continuing lack of diversity in environmental sociology (Mascarenhas et al. 2016). It finds that environmental sociology is still overwhelmingly white, that African American and Latine scholars are very much underrepresented in the section, and that women make up just 40 percent of regular members (excluding students). The study's authors point out that this lack of diversity restricts the range of topics, skills, and insights covered in the field. Moreover, a 2021 follow-up report by the section's membership committee analyzed changes in section membership between the years 2017 and 2020 and found that "scholars of color, particularly African Americans and Native Americans, remain significantly underrepresented" (Liévanos 2021). In this sense, environmental sociology and mainstream environmental movements, as Taylor (2016) shows, both have historically reflected broader racism in US society. By reckoning with this history of exclusion, this volume is, indeed, seeking to promote principles of racial integration in the production and value of knowledge.

A broad diversity of standpoints is particularly crucial in a field like environmental sociology, one where intersections of race, class, and gender shape outcomes to a great extent. Yet, in large part because of this lack of diversity, perspectives relevant to marginalized and oppressed groups, crucial to understanding human-environment relations, are more likely to remain at the edges of the field. Despite this, it is these very perspectives that have played a transformative role in environmental sociology in recent decades: Black scholars have been the first to develop a conception of environmental racism, whereby the distribution of environmental impacts varies according to race (Bullard 1990; Chavis 1987; Taylor 1992). This line of thinking was foundational in developing contemporary understandings of environmental justice—now a field of research and a social movement exposing injustices and demanding that all people are equally protected from environmental harm and have equal access to environmental benefits. In the same vein, many women scholars shed light not only on the gendered character of environmental harm but also on links between patriarchy and the destruction of nature (Wynter 2003). Finally, Indigenous resurgence has opened up a new set of perspectives on decolonization and relation to the land that slowly make their way into environmental sociology, complicating notions of environmental justice (McKay et al. 2020).

Diversity is thus redefining the work of environmental sociologists and their assumptions about the world. If, for example, as Ariel Salleh (2017) argues, part of our work is to challenge the fundamental Western belief that the nature of men is to dominate both women and nature, then intersectional approaches have relevance for environmental sociologists just as much as for gender and race scholars. Intersectional work and a diversity of scholars are greatly needed in environmental sociology—hence this volume's contribution to foster a broad range of standpoints and perspectives.

JUSTICE

The issue of justice is increasingly central to environmental issues. While justice was never fully absent from environmental sociological conversations, an emphasis on justice represents a departure from the uneasy relationship that many environmental justice scholars have had with environmental sociologists, an uneasiness that centered on disagreements over where any disciplinary or other boundaries might or should be drawn (e.g., Liévanos et al. 2021). However, such lines now tend to be blurred or ignored as concerns over justice guide research questions and agendas in environmental sociology (Harrison 2019; Pellow and Brehm 2013).

Other sociological subfields have grappled with the tension between value-free and value-laden research. For instance, leading race and intersectionality scholar Tanya Golash-Boza (2016) is unequivocal in her stance: "the study of race must be political and politicized because there is no good reason to study race other than working toward the elimination of racial oppression" (130). Such a statement has parallel concerns for environmental studies and should provoke self-reflection among environmental sociologists, who assess how to integrate moral values into scientific pursuits. In particular, now that justice is becoming a central concern for all, it is crucial to ask how we can better tie justice into our conversations regarding health, toxicity, decolonialism, climate change, and agriculture, among other issues. By pursuing such justice-oriented research, environmental sociology could become a model for emancipatory sociology (Morris 2017).

Diversifying the knowledge and experience of what constitutes environmental sociology can enrich the study and implementation of justice. In scholarship outside the United States, we see vivid examples of intersectional approaches to environmental justice. Afro-Brazilian historian Beatriz Nascimento (2023) has drawn from the experiences of Black women leadership in *quilombos*, communities traditionally populated by descendants of escaped enslaved people, to develop feminist theories and methods related to land rights, autonomy, and environmental justice. Farhana Sultana (2020) analyzes urban Bangladesh to show how the entanglements between gender, citizenship, and class shape unequal access to water. The combination of patriarchal expectations, low-income status, and legal marginalization force many urban women to do the daily work of retrieving clean water. This scholarship underscores critiques (see Sze 2017) that early environmental justice debates used excessively narrow parameters, often treating race and class as separate categories or conceptualizing gender as if it were a nonracial social construction. Intersectional scholarship, particularly in international contexts, can bring innovative insights to environmental sociology by highlighting the myriad axes through which environmental injustice and justice can occur.

CREATIVITY

One of the first steps in learning how to take part in an academic conversation, or any conversation for that matter, is developing an understanding of that conversation's language. By language we do not mean something like Cantonese or French or Urdu but rather a collection of things like the terms and concepts that are often used to make a point, the formats through which arguments are expected to be made, the various histories that inform current thinking, and the power dynamics that influence dialogue. Without an understanding of this language, a conversation—especially an academic conversation—can appear impenetrable. Once this language is learned, though, even partially, what was once incomprehensible can become crystal clear.

Sometimes, however, languages can calcify. They become so habitual that deviating from the established formats, concepts, histories, or power dynamics is an act of deviating from the conversation itself. The worst-case scenario is when the mere use of an established language is mistaken for a quality contribution, a mistake that usually only serves to further the interests of an entrenched hierarchy as opposed to the conversation itself. Thankfully, this worst case is not the case with contemporary environmental sociology. Nevertheless, we feel that it is important to time and again remind ourselves that the way our current language functions is not set in stone. We can embrace different formats, histories, concepts, and power dynamics in order to further environmental sociology in ways that sticking to our present language cannot. For this reason, from our initial inquiries forward we have emphasized how every author is free to buck routinized formats as they see fit. Creativity has been encouraged. We have told authors, "don't be boring."

This emphasis, however, was always an opportunity, not a demand. There are many good reasons to retain aspects of a common language. Whether and how to deviate from them is a delicate balance, and one that is different for everyone. While no author submitted an essay that leans on, for example, abstract poetry to make its point, each author has found a way to be creative in their own manner. The result is a collection of essays that not only reassesses where environmental sociology might go but rethinks how it might operate.

TEACHING AND RESEARCH

Perhaps the key factor in keeping any field fresh is consistent dialogue with incoming students. This is because those that are new to the field are the ones who are most likely to present new ideas and ways of thinking that question its trajectory. While no author in this volume is brand-new to environmental sociology, every chapter was written by an emerging scholar (or two or three emerging scholars) who has reflected on their teaching and research practice in order to bring a fresh idea or two to the proverbial environmental sociology table. Of course, no one's goal is parricide. Rather, the goal is to build through the discipline's history but not be bound by it. As a result, these emerging scholars concretize some novel ways of thinking so that they can in turn be taught to the next generation of scholars, and then, hopefully, be challenged anew.

This volume is geared toward teaching. It is not, however, intended to replace older texts, for these provide the foundations of the perspectives developed in each chapter. Our aim is rather to make available current dialogues within environmental sociology. Indeed, the texts in this volume form part of an ongoing conversation with the "canonical" literature used in teaching. In addition, they also engage new traditions that we think are important for our current era.

Scholars both within and outside environmental sociology will also find interest in these essays. The ideas developed will be of use to those working on closely related topics, not because they offer a comprehensive review of current thinking on the topic—which they intentionally do not—but, on the contrary, because they provide original viewpoints on how a discussion should move forward. They provide food for thought, and they sometimes even launch a critique. This volume is also ideal for environmental scholars who do not identify as environmental sociologists. It will provide creative ways of thinking through the foundations of how environmental and social processes combine, while also being open to whatever the future may bring. Our hope is that this volume will inspire fruitful discussions in many disciplines.

MOVING FORWARD

As editors, we did not approach founding figures or well-established scholars to make contributions, but rather their students, as well as some from outside these networks. We made an explicit attempt to include emerging scholars of diverse theoretical, institutional, gender identity, and racial backgrounds. In addition, authors have been discouraged from providing an "overview" of a particular conversation but encouraged to provide a creative argument for where a conversation is going.

The fourteen chapters that make up the rest of this volume engage a wide range of subareas in environmental sociology, if they do not necessarily provide an exhaustive overview of the entire subdiscipline. Nevertheless, each chapter represents an issue at the core of environmental sociology in North America, rendering this collection a launching pad for future debates, even for topics not included in this volume.

These chapters are divided into three sections. The first section, "Moving Within and Beyond the Discipline," provides multiple pathways for deeper disciplinary and interdisciplinary engagement. On the one hand, environmental sociologists are well-positioned to make profound and lasting contributions within the discipline on topics that are long-established in other fields but relatively new to sociology. On the other hand, this section illustrates the necessity to look beyond disciplinary limits and to engage in interdisciplinary collaboration. Several contributing authors have a deep, long-lasting commitment to interdisciplinarity through research, teaching, and professional service. Their approach to environmental sociology therefore often extends beyond conventional sociological boundaries. They challenge sociologists to consider arguments and evidence from other social and environmental scientists, while also creating opportunities to integrate sociological expertise and analysis into other fields.

Jordan Fox, Patrick Greiner, and Daniel Shtob begin the volume by cutting through the cacophony of voices that argue over the definition of, theoretical models of, and methods to study our "environmental" relationships. Instead of demanding that all environmental sociologists employ a singular approach to what the word "environmental" means and how to use it, they argue that environmental sociologists must first have the humility to acknowledge that our environmental and social relationships are so messy, so contextual, so complex, and so multifaceted that no one approach can explain them alone. We therefore need multiple ways of knowing and multiple ways of understanding the boundaries of what may or may not be "environmental." Still, this should not mean that "anything goes," they claim, but instead that we should develop a more attuned awareness of each approach's strengths and weaknesses, employing them according to their likelihood of generating successful explanations.

In a lively essay, Matthew Houser draws on a mix of existing literature as well as his own experience as an applied, transdisciplinary, environmental sociologist to uncover the general importance of the transdisciplinary turn, or the turn toward work that intentionally integrates research from different disciplines to a shared goal that incorporates stakeholder input into research design and focuses on addressing real world problems that have an impact on people's lives. He then outlines how environmental sociologists have the tools to, at once, engage in core sociological activities while also conducting applied, collaborative, transdisciplinary work. This leads into a discussion of both specific constraints and more structural factors built into the very fabric of the modern university that today create barriers to transdisciplinarity, as well as some means of overcoming them.

Cameron T. Whitley and Abraham Vanselow's essay is a consideration of animals. They argue that our relationships with animals are deeply embedded in systems of power and domination. Consequently, they argue, how we choose to engage the inhabitants of those natural environments—human or nonhuman—has profound implications for our shared worlds. Nevertheless, sociologists have historically ignored the human-animal connection, instead choosing to focus on the ways in which animals have been socially constructed and forgoing an examination of their material realities. Furthermore, they contend, while the popularity of environmental sociology has increased, animals are often reduced to being just another part of nature. They consequently advocate for a better integration of sociological animal studies and environmental sociology as a logical step toward a more inclusive examination of the connections between society, animals, and the environment. Applying queer theory, they suggest a polysemic approach to the animalizing of environmental sociology, or an approach that asserts a recognition of the multiple roles the researcher plays as scholar and activist as well as the multiple roles the animal plays as subject, entertainer, companion, food, and so on, and how the human and nonhuman can be conspirators in addressing environmental ills.

Alissa Cordner concludes this section with an essay considering environmental sociology's unique position as a subdiscipline that is not only capable of contributing to scholarly understanding of environmental issues but to meaningful and justice-oriented action on pressing environmental problems. She presents a model of engaged public sociology, one that allows researchers to collaborate with impacted communities and scholars across many disciplines to develop knowledge addressing research topics of interest to those most affected by environmental issues. Using case studies ranging from climate change and fossil fuel extraction to the health impacts of toxic chemicals, she identifies the challenges and opportunities of this model of engaged public sociology. In particular, she highlights how environmental sociology can produce knowledge and contribute to future action at multiple scales to understand and address long-standing environmental injustices and to improve public health.

Part of the foundational ethos of this volume is that authors should be open to respectful disagreement. For, like any way of knowing, environmental sociology is not a collection of functional pieces that make up some uniform whole, but a messy, sometimes discordant, conversation. And everyone involved in this project thinks that it is necessary to describe environmental sociology accurately, divisions and all. Yet one environmental sociological principle we all agree on, whether or not we agree on anything else, is that any conclusion that promotes injustice is fundamentally wrong. This is why the chapters in the second section of this volume, "Environment, Inequality, and Justice," place justice at their center. This means historicizing environmental degradation to demonstrate how many of today's environmental problems were built through ongoing structures of racial capitalism, patriarchy, and settler colonialism. It also means questioning the modernist narratives that serve to justify a present social order, one that is only barely coming to grips with patriarchal and otherwise intersectional forms of environmental injustice. Each chapter in this section takes on these challenges.

Ian Carrillo begins this second section by detailing the utility of the racial capitalism approach for advancing debates that are central to the environmental sociology canon. He does so by selectively drawing from foundational racial capitalism texts and other environmental sociology scholarship, using them to examine relationships between racial capitalism and the environment at each of the macro-, meso-, and micro-scales. He then concludes the chapter by discussing how advancing the study of racial capitalism and the environment necessitates not only the utilization of disciplinary and inter-disciplinary approaches but also attention to the development of theory and praxis oriented around total liberation.

From, of all places, a reference to David Fincher's 1999 film *Fight Club,* Jules Bacon and Kirsten Vinyeta seek to render settler colonialism more visible in environmental sociology (as well as sociology more generally). They consider the wide variety of ways in which settler colonialism structures environmental practices, policies, and epistemologies among non-Natives. They argue that through an analysis of the practices of state and corporate actors, as well as the cultural practices of mainstream environmentalism, we can better understand how settler-colonial ideologies inform a wide range of eco-social behaviors that we might consider both pro- and anti-environmental. They show how these behaviors tend to share common elements of Indigenous erasure and misinformation about Indigenous peoples and result in outcomes that negatively impact Indigenous peoples and communities. In response to this, in turn, Indigenous peoples mount unique forms of resistance aimed at defending their lands, waters, and cultures.

Michael Warren Murphy's essay runs against the latent colonial currents of hegemonic social science, pointing to the importance of an anticolonial approach to environmental sociological inquiry. He argues that our apprehension of the relationship between society and environment has often been limited by the occlusion of empire and colonialism. In addition, Murphy explores the theoretical assumptions that undergird an anticolonial approach to environmental sociology, highlighting existing research that exemplifies the promise and potential of such an approach.

Christine Labuski and Shannon Elizabeth Bell's essay is a critical examination of some of the central claims of ecofeminism, yet also an argument that ecofeminism offers necessary insights for radical environmental thinking, particularly given its robust analyses of power and human supremacy. The authors maintain that ecofeminism's attention to how power relations structure society's relationship with the more-than-human world makes it particularly valuable for understanding the intertwined ecological and social crises of our time. Moreover, they show, ecofeminism offers valuable lessons that can help us better imagine environmental futures that are racially and economically just, promote multispecies flourishing, value disabled, queer, and Indigenous lives, and acknowledge the importance of socially reproductive—and feminized—forms of labor.

Tanesha Thomas begins her essay with a discussion of how participants within the environmental justice movement have always been from diverse backgrounds. Although these Navajo miners, Latina mothers, Black farmers, and white working-class housewives have not only defended their families and communities against incinerators, pesticides, and other toxic hazards, they have also long been aware of the overlapping and intersecting nature of their own histories. With this context as a starting point, her essay discusses relationships between the environmental justice movement and intersectionality, highlighting the ways that contemporary researchers and activists are pushing the boundaries of traditional methodologies and explicitly showing how intersectionality and environmental justice impact the lives and ecosystems around us.

This volume concludes with a suite of chapters that together ask the question of "Transformations." As these chapters show, we should not be fooled into thinking that *anything* is possible. Rather, what our environmental relationships look like in the near, or even distant, future very much depends on contemporary power structures, ideology, politics, technologies, and social movements. Moreover, each of these influences will impact—for good or bad—different people in different ways. Indeed, as the chapters in this section demonstrate, social inequalities are a part and parcel of ecological relationships.

J.P. Sapinski opens this final section with a consideration of the corporation. There seems to be plenty of agreement that environmental sociologists ought to consider corporations in their work. Yet, he asks, is the corporation, as a social institution, really a core element of environmental sociological analysis? He argues that the corporation is seldom addressed, in itself, as an overwhelming influence. To remedy this, he draws from environmental sociology and economic sociology to outline the corporation as *the* key institution at the interface between humans and their environment in capitalist societies.

In a chapter that will surely instigate debate, Holly Jean Buck discusses a huge problem, one that is not so much technological but rather about our approach to technology. According to Buck, the problem is this: Environmental sociologists and related thinkers are stuck in a paradigm of instinctively critiquing technological development. Yet navigating environmental crises—such as climate change, water scarcity, or waste—requires the design and use of a variety of new and existing technologies deployed at planetary scales. We therefore need, she argues, to get over our fear of technology and think critically about how we can build an environmentalism that can not only critique but *guide* the development of all kinds of technologies and practices, from the personal to the industrial.

Amalia Leguizamón discusses the violence and injustice of the contemporary global food system. Her chapter begins by telling the story of how we got here—the conventional, globalized, corporate food system—through the expansion of agrarian frontiers for capitalist accumulation. Her goal is to highlight the power dynamics and ideologies that lead us to where we should assess the likely successes of two alternative futures already underway. One, proposed by ecological modernization scholars and corporate leaders alike, focuses on technological innovations like robots and genetically modified crops. The other fundamentally challenges the cheapening of lives and nature of corporate-led, tech-based agriculture by empowering peasants and shortening the distance between producers and consumers. This alternative centers on justice, agroecological methods, and Indigenous ways of knowing and being in the world.

Hillary Angelo's essay asks how the environment figures into contemporary urban analysis. Her chapter treats climate crises as opportunities for reflection on one historical point of intersection between urbanization and the environment: the "imperialism" of cities and metropoles over regional and global hinterlands. Angelo argues that while certain features of contemporary socio-environmental problems are new, these basic, persistent, material, and ideological relations are not. The chapter tells a brief history of urban imperialism through the growth of Western cities in North America, highlighting the political and material asymmetries that have characterized capitalist urbanization in the past two centuries. In the context of climate change, Angelo's reflections invite readers to reconsider these dynamics and what opportunities exist to transform them.

Diana Stuart closes the volume by discussing how scholars and activists in specific places are taking measures to address environmental crises. There are clearly competing worldviews and ideas about how to address our existential environmental threats and, based on the evidence, it is clear that some paths are much riskier than others. In addition, different sets of winners and losers will emerge depending on what solution pathways are chosen by world leaders, and what vested interests are actively working to maintain current power relations. In an age of global crises, it is easy to feel powerless, doomed, and hopeless. As the book's final chapter, this essay concludes by discussing how defeatism and fatalism only undermine possibilities for a more sustainable and just future and how such sentiments are already being propagated as a strategy to maintain the status quo.

Above all, we hope that these chapters demonstrate the breadth and scope of sociological inquiry on environmental and climate issues while also showing that the sociological study of the environment is needed across the many subfields of the discipline and beyond. The contributing authors not only advance the field of environmental sociology but also use the study of the environment and climate to push the limits of sociological research. We look forward to the conversations, debates, disagreements, and collaborations that emerge from the chapters included in *Environmental Sociology Now*.

NOTE

1. For instance, see Lidskog et al. 2015 on the differences between environmental sociology in the United States and in Europe; Hasegawa 2021 and Zinda et al. 2018 for overviews of environmental sociology in Japan and China, respectively; and Cock 2023 for a discussion of how to further environmental sociology in South Africa.

REFERENCES

Auerbach, Daniel, and Brett Clark. 2018. "Metabolic Rifts, Temporal Imperatives, and Geographical Shifts: Logging in the Adirondack Forest in the 1800s." *International Critical Thought* 8 (3): 468–86.

Baker, Zeke. 2024. *Governing the Climate: How Science and Politics Have Shaped Our Environmental Future*. University of California Press.

Besek, Jordan F., and Richard York. 2019. "Toward a Sociology of Biodiversity Loss." *Social Currents* 6 (3): 239–54.

Bullard, Robert. 1990. *Dumping in Dixie*. Westview Press.

Caniglia, Beth S., Andrew Jorgenson, Stephanie A. Malin, Lori Peek, and David N. Pellow. 2021. "Introduction: A Twenty-First Century Public Environmental Sociology." In *Handbook of Environmental Sociology*, edited by Beth Schaefer Caniglia, Andrew Jorgenson, Stephanie A. Malin, Lori Peek, David N. Pellow, and Xiaorui Huang. Springer.

Catton, William R., Jr., and Riley E. Dunlap. 1978. "Environmental Sociology: A New Paradigm." *American Sociologist* 13 (1): 41–49.

Chavis, Benjamin. 1987. *Toxic Wastes and Race in the United States*. United Church of Christ.

Cock, Jacklyn. 2023. "The Relation Between Sociology and Threats to Our Survival." *South African Review of Sociology* 53 (3): 249–58.

Dryden, P.N., and A. McCumber. 2017. "#Nature: Postmodern Narrative, Place, and Nature in Santa Barbara, CA." *Environmental Sociology* 3 (3): 286–96.

Ducre, Kishi. 2018. "The Black Feminist Spatial Imagination and an Intersectional Environmental Justice." *Environmental Sociology* 4 (1): 22–35.

Dunlap, Riley E., and Robert J. Brulle, eds. 2015. *Climate Change and Society: Sociological Perspectives*. Oxford Academic.

Ergas, Christina. 2021. *Surviving Collapse: Building Community Toward Radical Sustainability*. Oxford University Press.

Golash-Boza, Tanya. 2016. "A Critical and Comprehensive Sociological Theory of Race and Racism." *Sociology of Race and Ethnicity* 2 (2): 129–41.

Gould, Kenneth, and Tammy Lewis. 2016. *Green Gentrification: Urban Sustainability and the Struggle for Environmental Justice*. Routledge.

Grant, Don, Andrew Jorgenson, and Wesley Longhofer. 2020. *Super Polluters: Tackling the World's Largest Sites of Climate-Disrupting Emissions*. Columbia University Press.

Greiner, Patrick Trent. 2022."Colonial Contexts and the Feasibility of Mitigation through Transition: A Study of the Impact of Historical Processes on the Emissions Dynamics of Nation-States." *Global Environmental Change* 77:102609.

Gutiérrez-Zamora, Violeta, Imrelo Mustalahti, and Diego García-Osorio. 2022. "Plural Values of Forests and the Formation of Collective Capabilities: Learnings from Mexico's Community Forestry." *Environmental Sociology* 9 (2): 117–35.

Hasegawa, Koichi. 2021. "Japanese Environmental Sociology: Focus and Issues in Three Stages of Development." *International Sociology* 36 (2): 289–301.

Harrison, Jill Lindsey. 2019. *From the Inside Out*. MIT Press.

Holleman, Hannah. 2021. "Classical Theory and Environmental Sociology: Toward Deeper and Stronger Roots." In *The Cambridge Handbook of Environmental Sociology*, edited by Katharine Legun, Julie Keller, Michael Bell, and Michael Carolan. Cambridge University Press.

Kasper, Debbie. 2016. "Re-Conceptualizing (Environmental) Sociology." *Environmental Sociology* 2 (4): 322–32.

Leguizamón, Amalia. 2019. "The Gendered Dimensions of Resource Extractivism in Argentina's Soy Boom." *Latin American Perspectives* 46 (2): 199–216.

Liévanos, Raoul S. 2021. "SES Member Chair Memo: Recent Trends in SES Membership and Racial Composition." *Membership Committee Report to the American Sociological Association Environmental Sociology Section*. Section on Environmental Sociology, American Sociological Association.

Liévanos, Raoul S., Elisabeth Wilder, Lauren Richter, Jennifer Carrera, and Michael Mascarenhas. 2021. "Challenging the White Space of Environmental Sociology." *Environmental Sociology* 7 (2):103–9.

Lidskog, Rolf, Arthur P.J. Mol, and Peter Oosterveer. 2015. "Towards a Global Environmental Sociology? Legacies, Trends and Future Directions." *Current Sociology* 63 (3): 339–68.

Lidskog, Rolf, and Claire Waterton. 2016. "Conceptual Innovation in Environmental Sociology." *Environmental Sociology* 2 (4): 307–11.

Luna, Jessie K. 2017. "Getting Out of the Dirt: Racialized Modernity and Environmental Inequality in the Cotton Sector of Burkina Faso." *Environmental Sociology* 4 (2): 221–34.

Mascarenhas, Michael, Jennifer Carrera, Lauren Richter, and Elisabeth Wilder. 2016. *Diversity in Sociology and Environmental Sociology: What We Know About Our Discipline*. Section on Environmental Sociology, American Sociological Association.

McKay, Dwanna, Kirsten Vinyeta, and Kari Norgaard. 2020. "Theorizing Race and Settler Colonialism within U.S. Sociology." *Sociology Compass* 14 (9): e12821

Morris, Aldon. 2017. "The State of Sociology: The Case for Systemic Change." *Social Problems* 64 (2): 206–11.

Nascimento, Beatriz. 2023. *The Dialectic Is in the Sea*. University Press.

Pellow, David. 2017. *What is Critical Environmental Justice?* Polity.

Pellow, David, and Hollie Nyseth Brehm. 2013. "An Environmental Sociology for the Twenty-First Century." *Annual Review of Sociology* 39:229–50.

Porcelli, Apollonya, and Jordan Fox Besek. 2022. "Sub-Disciplining Science in Sociology: Bridges and Barriers Between STS and Environmental Sociology." *Environmental Sociology* 8 (2): 149–60.

Rath, Swatiprava, and Pranay Kumar Swain. 2024. "Investigating waste in the ambit of environmental sociology in Bhubaneswar, India." *Environmental Sociology* 10 (2): 192–205.

Rea, C.M., and Frickel, S. 2023. "The Environmental State: Nature and the Politics of Environmental Protection." *Sociological Theory* 41 (3): 255–81.
Rudel, Thomas K. 2009. "How Do People Transform Landscapes? A Sociological Perspective on Suburban Sprawl and Tropical Deforestation." *American Journal of Sociology* 115 (1): 129–54.
Salleh, Ariel. 2017. *Ecofeminism as Politics: Nature, Marx and the Postmodern.* Zed Books.
Scoville, Caleb. 2022. "Constructing Environmental Compliance: Law, Science, and Endangered Species Conservation in California's Delta." *American Journal of Sociology* 127 (4): 1094–1150.
Steinmetz, George. 2005. "Introduction: Positivism and Its Others in the Social Sciences." In *The Politics of Method in the Human Sciences: Positivism and Its Epistemological Others*, edited by George Steinmetz. Duke University Press.
Stuart, Diana. 2016. "Crossing the 'Great Divide' in Practice: Theoretical Approaches for Sociology in Interdisciplinary Environmental Research." *Environmental Sociology* 2 (2): 118–31.
Sultana, Farhana. 2020. "Embodied Intersectionalities of Urban Citizenship: Water, Infrastructure, and Gender in the Global South." *Annals of the American Association of Geographers* 110 (5): 1407–24.
Sze, Julie. 2017. "Gender and Environmental Justice." In *The Routledge Handbook of Gender and Environment*, edited by Sherilyn MacGregor. Routledge.
Theis, Nicholas. 2020. "The Global Trade in E-Waste: A Network Approach." *Environmental Sociology* 7 (1): 76–89.
Taylor, Dorceta. 1992. "The Environmental Justice Movement." *EPA Journal* 18 (1): 23.
Taylor, Dorceta. 2016. *The Rise of the American Conservation Movement.* Duke University Press.
Wynter, Sylvia. 2003. "Unsettling the Coloniality of Being/Power/Truth/Freedom: Towards the Human, After Man, Its Overrepresentation—An Argument." *CR: The New Centennial Review* 3 (3): 257–337.
Zinda, John Aloysius, Yifei Li, and John Chung-En Liu. 2018. "China's Summons for Environmental Sociology." *Current Sociology* 66 (6): 867–85.

PART I

Moving Within and Beyond the Discipline

1. What's So Environmental about Environmental Sociology?

Jordan Fox, Patrick Trent Greiner, and Daniel A. Shtob

It should come as little surprise that what distinguishes environmental sociologists from other sorts of sociologists is the fact that we are concerned with processes that are environmental. We, as one might expect, consequently place a tremendous amount of importance on this word: "**environment**." Its presence is our justification for existing as a distinct sociological community. It is what inspires our work. It is—in large part—what distinguishes our intellectual identity.

But what does "environment" mean here, exactly? At first blush, the answer appears self-evident. If a process is environmental, it has to do with the entirety of the "natural world," and all the forests and migrations and climates and beetles and other animals and plants and systems contained therein. When environmental sociologists use the word environmental, it is true that we are usually referring to these sorts of processes.

Though, typically, to say a process is environmental does not only mean it has to do with the *presence* of wolves and seeds and mountains and tides and other aspects of the natural world but also with the *absence* of society. To call a process environmental is often to contrast it to the cities and technologies and inequalities that, for better or worse, seem to characterize the social dynamics of our species. To call a process environmental, in other words, is often to signify the absence of human beings.

It may complicate things a bit, then, that when environmental sociologists use the word "environmental," this is not what we mean.

For us, environmental processes actually include a great deal of human societies and relationships. Like all organisms, human beings have ecological, biological, and other sorts of material relationships that enable or constrain our collective action. A moment's reflection on the social and ecological relationships involved during your last trip to the grocery store, visit to the hospital, or stop to fill a car's gas tank will render some of these clear. It should also be clear that our environmental relationships are extremely variable, and often in ways that reflect inequality and difference. For instance, race and racism shape our environmental relationships, but do not necessarily do so in the same ways in Brazil as they do in the United States (Carrillo 2021). Environmental sociology is indeed not concerned with how society in general interacts with a single, stable "environment," but rather with how specific social relationships—for example, those that have to do with gender (e.g., Kennedy and Dzialo 2015) or colonialism (e.g., Murphy 2020; Greiner 2022) or social movements (e.g., Kallman and Frickel 2018)—interact with and coproduce specific environmental processes. Relationships between social and environmental processes are always shifting, always interacting at multiple scales, always transforming. Sometimes these changes are extremely slow and seemingly uneventful (Ollinaho 2015)—for instance, the gradual accumulation of greenhouse gasses in the atmosphere. Sometimes they are immediate—for instance, when a mining company blows off the top of a mountain to access the coal within.

Now, let us ask the question from the second paragraph of this chapter in another way: For an environmental sociologist, what is *not* environmental? Finding a satisfying answer to this question can become extremely frustrating. For how might we draw a line between, for example, capitalism as an environmental process and capitalism as a social process? Can there be a line? Should there be a line? Or is it necessary to problematize any and all distinctions between environmental and capitalist processes, no matter what? Some environmental sociologists argue that a nuanced separation is an analytic necessity (e.g., Carolan 2005; Foster 2016), while others are insistent that any separation at all is unacceptable (e.g., White, Rudy, and Gareau 2015).

Moreover, *how* we answer these questions has important implications for our relationships with each other. For instance, if we decide to reject any meaningful distinction between environment and society, should we then reject the work of those who accept such distinctions? If so, environmental sociologists would be forced to reject the research of most other sociologists, a majority of whom separate out environmental processes from their work. In addition, environmental sociologists would be forced to reject the research of most natural scientists, a majority of whom do not incorporate social processes. In all, rejecting the research of scholars who either implicitly or explicitly separate environment and society would mean refusing to include much of our own discipline of sociology, as well as much of the scientific work at the root of our understandings of climate change, biodiversity loss, and much else of fundamental concern to environmental sociologists. On the other hand, if we decide to draw a sharp line between what is environmental and what is not, should we reject those Indigenous ways of knowing in which the very idea that environment can be separated from "society" is seen as a step toward the destruction of Indigenous lifeways and livelihoods?

It would be a gross mischaracterization of environmental sociology to say that we all agree on answers to these questions, despite the overwhelming importance of the signifier "environmental" to our subdiscipline. There are many active environmental sociological traditions, each with differing relationships to Indigenous thought, natural science, race, capitalism, and the broader discipline of sociology (Stuart 2016). What ultimately makes environmental sociology "environmental" for each of us is often quite different.

In this chapter, we will argue that such disagreement over what is or is not "environmental" is not necessarily a problem for environmental sociology. If embraced, it can be a strength.

Considering the overwhelming **complexity** of our environmental relationships, we think it would be detrimental to environmental sociology if we (or anyone) demanded that all environmental sociologists employ a singular approach to what the word "environmental" means and how to use it. Instead, we need to develop the humility to embrace environmental complexity, to understand that we are error-prone scholars attempting to cull a sense of stability from messy, disjointed natural worlds, and sometimes we need more than one way of demarking and approaching environmental questions. In brief, we need to embrace a more **philosophically pluralist** environmental sociology, accepting that multiple principles and ideas can coexist.

Environmental sociology will benefit from intentionally cultivating multiple ways of knowing, each coming from different intellectual and material contexts, and each offering different strengths and weaknesses. Any single approach to demarcating the boundaries of what counts or what does not count as "environmental" would constrict our ability to confront the deep complexity of our world.

Both the environmental and social dynamics we explore constantly illustrate the inadequacies of our taxonomies and the illusiveness of "natural kinds." These dynamics, in their defiance of easy categorization, remind us that all approaches are limited and partial. They are all missing something (even if they offer many insights). We can and should, then, expand our ways of understanding to accommodate the fact that many of us have vastly different environmental sociological, scientific, political, and/or other aims that require vastly different pursuits of knowledge. The world does not abstract and order its problems for us into neat and separable boxes of things that are environmental or social or socio-environmental. Instead, we abstract from the world ourselves, via our theoretical points of view and through the demands of our methodological approaches. Our abstractions may be powerful, and they may offer a way to reduce complex, often multicausal, real-world dynamics into frameworks that can be more easily discussed, analyzed, and communicated to nonexperts. In the course of our abstractions, however, we necessarily cut away much of the world. As powerful as our abstractions can be, they are all necessarily partial and in need of complement if we share a goal of understanding how, when, and where the social and environmental bear on one another (York and Clark 2010). We should encourage each other, then, to develop a plurality of theoretical points of view and methodological approaches that are each, in their own ways, successful at explaining what they want to explain.

In the following section we will explore the idea of "environment." This exploration will provide the foundation for our third section, where we develop our pluralist approach to environmental sociology. This approach is designed to be interdisciplinary, diverse, and justice-oriented, though it is also designed to avoid the traps of extreme scientism, by which we mean the idea that the world can only be understood and measured according to fixed, general, and necessary laws, as well as the traps of extreme relativism, by which we mean the idea that environmental ideas merely reflect the cultural viewpoint of their producer. Instead, ideas are judged based on whether or not they can bring the complexity, contingency, and deep uncertainty of social and environmental relationships to the forefront of analyses. Our hope is that such an approach can provide the basis for more productive, insightful, and respectful engagement among diverse environmental sociological traditions, as well as engagement between environmental sociology and other ways of understanding the world.

ON THE VERY IDEA OF ENVIRONMENTAL IDEAS

Once upon a time, environment was all that there was. While the earth itself formed around four and a half billion years ago, current estimates suggest that human history began a little more than two million years ago, while our species, homo sapiens have traced our evolutionary origins back a mere two to three hundred thousand years. This means that our species has been present for only about 0.0055 percent of the history of this planet.

Those processes we typically think of as environmental—those rains and insect migrations and river formations and microbial interactions—have mostly existed for far longer than we have. Though human societies have done a lot with this 0.0055 percent. Indeed, it is, in large part, *because* of how much we have done that environmental processes have changed much more than they otherwise would have over this short period.

Still, the mere fact that we change environmental processes does not make homo sapiens particularly remarkable or unique. All species, homo sapiens included, change environments just as they are changed by them. The evolutionary geneticist Richard Lewontin, for instance, has forcefully argued that "genes, organisms, and environments" are in constant, communal, interaction with each other (2002; also see Nicholson and Dupré 2018). One thing this constant interaction means is that, if we are to effectively think about our environmental relationships and how they influence outcomes of interest to us, we must avoid explanations that are *monocausal*—that is, we must avoid explanations that claim things are the way they are because a singular social or environmental process made them that way. Environments do not control societies, and societies do not control environments. Thinking we can discover some covering explanation that shows how one controls the other misses the point and diminishes the vast complexity and beautiful diversity of our world. It also encourages the development of a rationale that underestimates the scale of the analytical and political challenges environmental sociologists face.

Our field is still in its infancy, and intellectual hubris will not speed its development. Instead of searching for deterministic laws, we would be better served by focusing on how specific environments and specific social processes change through each other. We must focus on the complexity, contingency, and dynamism of our relationships. We must embrace the dizzying difficulty of our task and acknowledge just how little any one theoretical perspective or methodology can reveal on its own—powerful as it may be. In ways subtle and immense, in ways large and small, environments influence the development of social life just as societies create, modify, and make choices about the environments in which they live. The environmental idea that *everything is connected* is fundamentally true.

Still, affirming that everything is connected, in itself, doesn't help us produce better analyses of the world. While we should always keep ultimate connections in mind, environmental sociological analyses are about focusing on the dynamics that set specific social and environmental relationships in motion. Some of these dynamics may be intensely local. For instance, if an unelected, state-appointed manager switches the drinking water source of a majority Black city to a nearby river to save money, this local decision may combine with the chemical history of that aquatic ecosystem to produce drinking water that is hazardous to humans, causing a local health crisis. This is how the water supply of Flint, Michigan, was poisoned in the 2010s, creating one of our era's most significant environmental justice events. Nevertheless, a more comprehensive explanation of this event would not only focus on these local dynamics but also on how dynamics at larger historical, social, and environmental scales helped to structure it. Explanation would need to incorporate a strong understanding of race and racialization, the history of deindustrialization and racial formation in Michigan, the development of the American legal system, the voices of local residents, the ways in which capitalist imperatives tend to organize social and environmental relationships, the ecological history of the city, knowledge of water system infrastructures, a basic understanding of chemistry and epidemiology, the ways in which government agencies use chemistry and epidemiology to set the threshold for how much lead in drinking water is "acceptable" for human health, and much else.[1] An awareness of each of these social and environmental processes—their corresponding contexts, their various scales, their complexity, and the ways they influence and change each other—is essential.

Notably, the very idea that the English word "environment" has to do with the natural world at all originated from a similar insistence on this sort of interactive awareness. "Environment," which has been in use in the English language since the sixteenth century, is derived from the Old French verb *environner*, which more or less means "to surround." And "to surround" is generally what the word meant in English for hundreds of years. It still does mean this, of course, though it wasn't until nineteenth-century developments in what we would now call the philosophy of biology—specifically, novel conversations regarding how to best grasp the ways in which organisms relate to their surroundings—that the word "environment" took on the additional meaning of often having to do with nature (Pearce 2010; Sprenger et al. 2023).[2] Nevertheless, this newborn connection between the words "environment" and "nature" remained largely confined to more philosophical and scientific debates for quite some time.

There was no single twentieth-century event that caused the word "environment" to take on this additional, more natural meaning in the broader Western world. It was, rather, a series of events, all of which added to a growing awareness of the far-reaching influence of human beings on the planet's biophysical processes. Some of these events include: the spread of settler-colonial agriculture prompting a worldwide crises of soil erosion, culminating in "dust bowls" in the North American Midwest, South Africa, and inland Australia (Holleman 2018); how the acceleration of our nuclear capabilities provided select people the capacity for boundless destruction on a world scale (Arendt 2018; Rosa 2000); two world wars that decimated cultures, ecosystems, and ideologies across several continents; and imperialisms that decimated cultures, ecosystems, and ideologies across several more. There was a cold war in the wings, and much else besides. By the mid-twentieth century, it was clear that social and ecological worlds were changing at scales they had not before in human history, and it was rather unclear what the future held.

We needed to do a better job managing our far-reaching influence on the planet's biophysical processes. Part of this project was developing a new concept that could capture the scale, depth, and complexity of our open-ended relationships with nature. Throughout the 1940s, 50s, and 60s, an array of scholars, activists, policy-makers, and institutions would reformulate and then adopt the more technical, "natural," meaning of the word "environment" from biology to satisfy this need (Benson 2020; Warde, Robin, and Sörlin 2018).[3] The contemporary meaning of "environment" in popular discourse as a word that often has to do with nature is indeed the product of this particular time (the mid-twentieth century), locale (elite communities in North America and Western Europe), and a particular point of view (that we need to better manage our relationships with nature).

There has since been a lot of pressure on this word: "environment." As the 1960s rolled on, the word's association with the totality of nature moved beyond ivory towers, activist circles, and institutional boardrooms into broader public conversation. Its use has come to signify an integrated knowledge-making about the world, a belief that we can better understand and manage our relationships with nature, a politics that links local change to worldwide pressures. It has inspired new activist movements, new United Nations commitments, new art, and new academic fields (for example, environmental sociology).

Despite (or perhaps because of) its abundant usage, however, the precise meaning of the word "environment" remains rather fuzzy. One reason is that "environment" is a word that stands for an impossibly complex set of historical and contemporary phenomena—quite literally, the interactive totality of nature at all scales—but it reduces this deep complexity into a neat, manageable signifier. "Environment," as a result, stands for the profound and unwieldy intricacy of nature, though it does so without demanding any need for further explanation (Sprenger 2023, 411). Most of the time we use the word, we can just assume that environmental complexity is self-evident and simply move on with our conversation. In a sense, then, the word "environment" performs a magic trick. It accepts the complexity of nature and then makes that complexity disappear.

Such a magic trick can quickly become a problem. For instance, in the beginning of this section, we wrote that human societies have done a lot with our 0.0055 percent of the history of this planet, and in large part because of us environmental processes have changed dramatically over our time. This is true. Though speaking in such generalities about "environmental processes" and "our time" may make it seem as if all environmental processes are roughly the same and that all human beings are equally responsible for all environment change. And these things *are not* true. There are specific environmental processes that have changed, all in their own unique ways, and there are specific minorities of human beings that are responsible for a majority of environmental change. For example, members of the ownership groups that maintain our fossil fuel energy systems are far more responsible for climate change than oil rig workers (Hornborg and Malm 2014). If environmental sociology is to successfully explain social and environmental relationships, it is essential that we address the specifics of these dynamics. We have no choice but to dig into their complexity. It is more honest and, frankly, more exciting.

But how? To best do so, environmental sociologists must first have the humility to acknowledge that environmental and social relationships are so messy, so contextual, and so multifaceted that no one approach can achieve these goals alone. If we are to explain the dynamics we care most about, we need multiple ways of knowing and multiple ways of understanding the boundaries of what may or may not be environmental, from multiple standpoints. Still, we must be clear that not all approaches are appropriate at all times: each has their strengths and weaknesses. Consequently, we also need a way of evaluating the validity of our claims.

In the following section, we outline an approach to the word "environment" that embraces philosophical pluralism without abdicating judgment.

AN ENVIRONMENTAL SOCIOLOGY FOR LIMITED BEINGS

Environmental knowledge, like all knowledge, comes from experience. This is true no matter if you are a hydrogeologist trying to understand the processes that influence the fate of hazardous contaminants in groundwater, a member of the Maidu, Yuki, or Kawaiisu peoples learning to appropriately use the medicinal properties of valley oak tree bark, a ten-year-old following your older sister's instruction as you plant tomato starts, or an environmental sociologist analyzing a peasant farmer's stories about how multinational energy corporations have distorted local land dynamics. In each of these rather different cases, environmental knowledge is built the same way. Life experience and reflection on that experience is combined with previous knowledge to decide how to best engage with the world. Some of this knowledge comes from direct experience, while some comes secondhand. Nevertheless, in the process of building a socially organized body of knowledge, experience, and reflection on experience is always central.

Several important points follow. The first is that nearly all systems of environmental knowledge—whether from Western traditions or from non-Western, partially Western, and/or folk traditions—are built from a similar foundation. Different systems of knowledge, in other words, are typically quite conversant. Only in very rare cases or in very limited ways are they ever incommensurable. How else could so much of Western medicine have its roots in white settler contact with African American medicinal knowledge (which, itself, is likely in-part borrowed from Indigenous American healing techniques like those used by the Maidu, Yuki, or Kawaiisu)? If these systems of knowledge were fundamentally different, then Western pharmaceutical corporations would have little need to continue their exploitation of non-Western, partially Western, and/or folk medical traditions (Whyte 2018).

Second is that the practice of creating environmental knowledge is never solitary. It is instead the act of bringing together an extraordinary number of social and material processes and relations—technological, cultural, theoretical, physical, and more—into one's experiential arena to (hopefully) explain what one wants to explain and do what one wants to do. Creating environmental knowledge is about intertwining social and material history with local context and language, ceaselessly experimenting in an open-ended pursuit of explanatory success (cf. Pickering 1995; Rouse 2015).

Nevertheless, the world itself is active and changing. What worked last year may not work so well this year. Indeed, what all forms of environmental knowledge share is not only an origin in experience, reflection, and past knowledge but also the inevitability of being, at least to some extent, wrong.

Our third point starts from the fact that each of the examples from the first paragraph of this section—the hydrologist, the Maidu, Yuki, and Kawaiisu peoples, the ten-year-old, and the environmental sociologist—are divvying up and analyzing the world in quite different ways. Despite a shared origin in experience, no one form of environmental knowledge is exactly alike. Every history is different. Every context is different. Accordingly, all are likely to categorize and demarcate the world differently.

That different ways of knowing are, well, different should not come as a surprise. Too often, however, we feel the need to choose among different ways of approaching the word "environment." It is as if to take part in an environmental sociological conversation one must commit to either, say, feminist political ecology (Rocheleau 1996) or ecological Marxism (Burkett 2014), or between qualitative or quantitative methodologies, or between theories that retain distinctions between the environmental and the social (e.g., Carolan and Stuart 2015) and theories that claim all such separation is an inept exercise in "Cartesian" dualism (Moore 2011). This thinking suggests that if one approach to doing "environmental" work is correct, then the others must be altogether wrong. This is a false choice. Whether or not one categorizes a particular process as environmental has little to do with any capital "T" truth about the way the world is or is not. No one discovers what is or is not "environmental" or "social" or both "social and environmental" in nature. Instead, we demark reality along these lines based upon our theoretical point of view and, perhaps, our methodological limitations (Oreskes 1999). And, to this end, multiple theoretical points of view and multiple methodological approaches are potentially valuable in any given context. Insisting that only this or that perspective is the only "correct" environmental sociological perspective has more to do with branding one's work as a commodity for maximum consumption (or citation) and less to do with developing better ways to explain the world.

Our fourth point is that being open to multiple theoretical points of view does not mean being open to all theoretical points of view. Some theories are far more useful than others. For instance, "**scientistic**" theories are not useful. While there are many variants of scientism, in general a scientistic theory would treat relationships between social and environmental processes as if they could be understood simply by importing methods from the physical sciences. The reason this approach is not useful is because social and environmental relationships do not behave in the same way as many of the systems physical scientists developed their methodologies and theoretical frameworks to study. For example, an astronomer studying the movement of celestial bodies is able to predict, with stunning accuracy, what a planet's location will be millions of years in the future. In turn, an environmental sociologist studying the successes and failures of Canadian climate change activism would not be able to give a precise prediction of what such activism will look like even ten years from now. Social and environmental relationships do not behave according to fixed, discoverable, necessary, or predictable laws.[4] We should not pretend that they do.

Theories that fall under the umbrella of extreme **relativism** are also unhelpful, though for the opposite reason. Instead of overstressing how much we can know, these ways of thinking overstress how much we cannot know. For these ideas claim that because all theories are a product of human activity, they merely reflect the cultural viewpoint of their producer. Extreme relativism thus treats environmental knowledge as if it is created whole cloth from culture and is therefore unrelated to the world outside our heads. Yet if culture determines all, how could so many disparate cultures share so much environmental knowledge? How could, for example, so much of early Portuguese and Dutch botanical knowledge have been built through colonial contacts with lower-caste Southwest Indians? Indeed, early Portuguese and Dutch colonists so valued lower-caste knowledge precisely *because* it was built through the real-world collection and cultivation of plants across generations (Grove 1996, 87).[5] Certainly, culture plays a sizeable role in shaping the various constructions of environmental knowledge. As environmental sociologists, though, our task is not only to develop a sensitivity to cultural difference but also to develop a sensitivity to how much environmental experience we share.

The characteristic that useful theories share is not an ability to predict the future (scientism), nor is it a renouncement of all judgment (extreme relativism). What useful theories share is something much more realistic. They share an ability to help us define and measure the deep complexity and unpredictability of our environmental relationships. Including complexity in our analyses does not (or, rather, should not) entail hiding behind complex language or convoluted argument. Instead, it entails having the humility to be open to the fact that the world is so intricate, so dynamic, and so muddled that it needs to be understood in multiple ways. The way in which we understand social and environmental relationships should not be determined in advance but should focus on developing the best possible fit between our ways of understanding and the context we are trying to understand (Ermakoff 2017). Though, because no one theory can cover the deep complexity of social and environmental reality, multiple theories are needed. The point is for various environmental sociological theories, as well as theories as far afield as those in climatology and postcolonial studies, to complement each other in ways that can help guide our actions. The mature pluralist attitude is one that is not only open to engaging productively with approaches one does not fully agree with but is also capable of switching between different theories as needed in order to better understand and act in the world.

Note how throughout this section we have been referring to our "shared world." This is a very deliberate choice. For our perspective does not claim that there are multiple environmental worlds, each needing their own theory or methodology. It instead claims that, as Sandra Mitchell (2009) notes, "there are multiple correct and useful ways to describe the world" (14; see also Dupré 1999; Harding 2017). In other words, the environmental and social phenomena we are all trying to better understand are real, observable (even if only indirectly), and concrete. We share a material reality. Yet no one has a God's-eye view of our shared material reality. Our perspectives are always piecemeal, our taxonomies always a bit fuzzy, our knowledge always incomplete. By definition, our points of view are limited by our historical circumstance. Though these contextual limitations are not hurdles to one day overcome but reminders that we are real people pursuing knowledge in real situations. Just as the social and natural processes we study are deeply contextual, so is our knowledge of them. No one perspective on environment will be able to satisfy every environmental sociological need. Consequently, if we are to develop a better idea of the whole,—that is, the interactive, multiscalar complexity of our social and environmental relationships—we need to cultivate multiple systems of knowledge production, developing the sensitivity to incorporate the power and limits inherent in each.

Our fifth and final point is that our pluralism should cast a wide net. Some environmental sociologists, for instance, are deeply uncomfortable with relying on Western natural science in their work. We do not dispute the fact that there is good reason for this. For much of its history, Western natural science has not only served but has been built through capitalist, colonial, and imperialist aims (Grove 1996; Murphy 2021), while an overreliance on Western technological solutions to social and environmental problems has repeatedly exacerbated inequalities (Mansfield 2021; Pellow 2018), often rendering those very same social and environmental problems worse (York 2017; Holleman 2018; Levins 1996). In addition, our overreliance on Western natural science has, undoubtedly, exploited and then pushed aside non-Western, partially Western, and/or folk systems of knowledge.

Nevertheless, considering the urgency with which we must address climate change and the increasing dynamism of socio-environmental relationships, we cannot ignore the sheer number of efficient social and technological resources at the disposal of Western natural science. These resources, combined with a well-organized tradition of valuing critical reflection, provide Western natural science with tremendous problem-solving power that we would be foolish to ignore.[6] In other words, we should be careful not to essentialize our tools (Benjamin 2019). Moreover, much of Western natural science was produced as a means of a resistance to colonialism and other forms of exploitation, even for those in the metropole. It is not as if all

Western natural science was totally complicit with colonialism or other forms of exploitation, nor in the same ways (e.g., Costa 2023; Foster 2020; Levins and Lewontin 1984). Considering, for instance, the immensity of anticolonial art and literature derived from metropoles, we should know that exploitative ideologies were not all-encompassing (Steinmetz 2023). So, neither should our treatment of Western natural science.

In turn, some other environmental sociologists remain skeptical of non-Western, partially Western, and/or folk systems of knowledge. While few say it out loud, there is a long-standing sense that these systems of knowledge are not as rigorous, as generalizable, or as stable as Western systems of knowledge, and should therefore not be accorded the same respect. Yet, there are numerous instances in which systems of knowledge that derive from outside the Western canon have been far more successful, over the long run, than those that derive from within the Western canon. Think about the centuries-long successes of Indigenous controlled burning in California forests as opposed to the consistent failures of state management, or the long-term sustainability of using chemical fertilizers to grow food versus agroecological practices, many of which are derived from Indigenous knowledge. Even on Western terms, non-Western, partially Western, and/or folk systems of environmental knowledge are often—in the long term—far more successful at achieving what they want to achieve. Once we expand our time horizons beyond those demanded by the publishers, granting agencies, and tenure-committees of Western academia, there is a good case to be made that Indigenous theory on the whole has a tighter connection to empirics than does Western natural science (Anderson 2013).

If we are to develop better, justice-based solutions to the problems that define our socio-environmental present, we need to overcome any outright hostility to either of these ways of knowing. In fact, there is a convincing case to be made that the lines we often draw demarcating Western knowledge and other knowledges are far blurrier than most recognize, if indeed they are defensible at all (Graeber and Wengrow 2021). Moreover, the line between those environmental sociologists who defend an analytical or other sort of nuanced distinction between what is environmental and what is social and those who wish to dissolve such distinctions is also far less bright than most acknowledge. What counts or what does not count as "environmental" is not nearly as important as our ability to embrace the deep complexity, contingency, and unpredictability of our environmental relationships, building an environmental sociology that, through many levels and sites of analysis, comes to hold more than one point of view, and has the freedom to wield those perspectives as needed.

NOTES

1. For environmental sociological analyses of the poisoning of the Flint water supply, see Fasenfest 2019; Pulido 2016; and Shtob and Besek 2021.

2. The histories and alternative meanings of the word "nature" have been scrutinized far more extensively than those of the word "environment." For some well-known examples, see Cronon 1995; Soper 1995; Williams 1980.

3. The scholars, activists, policy-makers, institutions, and others that adopted the word "environment" were *far* from the first to examine our extensive relationships with nature or our capability for extensive transformations of nature. For instance, as far back as sixty thousand years ago, peoples in Africa used fire to manage ecological relationships, and, by the eighth century, various Persian, Afghani, and Northern Indian gardeners had developed botanical texts for managing gardens. In 1783, Hong Liangji wrote an essay warning China's Qing dynasty emperor of drastic, human-induced environmental change, while the German explorer and polymath Alexander von Humboldt sought to unify the sciences of his day to better understand anthropogenic climate change and other forms of human influence on nature (Burnside et al. 2023).

4. One should not take from this discussion that *all* natural scientific problems yield to the study of such laws. Evolutionary science, for instance, is so mired in contextual uncertainties that it tends to defy prediction. More contemporarily, a conservation ecologist interested in the migration patterns of mammalian fauna might also find traditional "scientism" of limited utility.

5. As opposed to Brahmin, or priestly caste, knowledge, which was merely composed from vague propositions and opinions from old texts.

6. Note how this argument does not rely on Western natural science being more "objective" or "universal" or reliant on a stable "scientific method." It instead relies on understanding the advantages and disadvantages of the way it is being produced.

REFERENCES

Anderson, M. Kat. 2013. *Tending the Wild: Native American Knowledge and the Management of California's Natural Resources.* University of California Press.

Arendt, Hannah. 2018. *The Human Condition.* 2nd ed. University of Chicago Press.

Benson, Etienne. 2020. *Surroundings: A History of Environments and Environmentalisms.* University of Chicago Press.

Burkett, Paul. 2014. Marx and Nature: A Red and Green Perspective. 2nd ed. Haymarket Books.

Burnside, William R., Simone Pulver, Kathryn J. Fiorella, Meghan L. Avolio, and Stephen M. Alexander, eds. 2023. *Foundations of Socio-Environmental Research: Legacy Readings with Commentaries.* Cambridge University Press.

Carolan, Michael. 2005. "Realism without Reductionism: Toward an Ecologically Embedded Sociology." *Human Ecology Review* 12 (1): 1–20.

Carolan, Michael, and Diana Stuart. 2016. "Get Real: Climate Change and All That 'It' Entails." *Sociologia Ruralis* 56 (1): 74–95.

Carrillo, Ian. 2021. "Racialized Organizations and Colorblind Racial Ideology in Brazil." *Sociology of Race and Ethnicity* 7 (1): 56–70.

Costa, James T. 2023. *Radical By Nature: The Revolutionary Life of Alfred Russel Wallace*. Princeton University Press.
Cronon, William. 1995. "The Trouble with Wilderness." In *Uncommon Ground: Toward Reinventing Nature*, edited by William Cronon. W.W. Norton.
Dupré, John. 1999. "Are Whales Fish?" In *Folkbiology*, edited by D.L. Medin and A. Atran. MIT Press.
Ermakoff, Ivan. 2017. "Shadow Plays: Theory's Perennial Challenges." *Sociological Theory* 35 (2): 128–37.
Fasenfest, David. 2019. "A Neoliberal Response to an Urban Crisis: Emergency Management in Flint, MI." *Critical Sociology* 45 (1): 33–47.
Foster, John Bellamy. 2016. "Marxism in the Anthropocene: Dialectical Rifts on the Left." *International Critical Thought* 6 (3): 1–29.
Foster, John Bellamy. 2020. *The Return of Nature: Socialism and Ecology*. Monthly Review Press.
Graeber, David, and David Wengrow. 2021. *The Dawn of Everything: A New History of Humanity*. Farrar, Straus, and Giroux.
Grove, Richard. 1995. *Green Imperialism: Colonial Expansion, Tropical Island Edens, and the Origin of Environmentalism, 1600–1860*. Cambridge University Press.
Harding, Sandra. 2017. "Democratizing Philosophy of Science for Local Knowledge Movements: Issues and Challenges." *Gender, Technology and Development* 4 (1): 1–23
Malm, Andreas, and Alf Hornborg. 2014. "The Geology of Mankind? A Critique of the Anthropocene Narrative." *Anthropocene Review* 1 (1): 62–69.
Holleman, Hannah. 2018. *Dust Bowls of Empire Imperialism, Environmental Politics, and the Injustice of "Green" Capitalism*. Yale University Press.
Kallman, Meghan Elizabeth and Scott Frickel. 2019. "Power to the People: Industrial Transition Movements and Energy Populism." *Environmental Sociology* 5 (3): 255–68.
Kennedy, Emily Huddart, and Liz Dzialo. 2015. "Locating Gender in Environmental Sociology." *Sociology Compass* 9 (10): 920–29.
Levins, Richard. 1996. "Ten Propositions on Science and Antiscience." *Social Text* 46/47:101–11.
Levins, Richard, and Richard Lewontin. 1984. *The Dialectical Biologist*. Harvard University Press.
Lewontin, Richard. 2002. *The Triple Helix: Gene, Organism, and Environment*. Harvard University Press.
Mansfield, Becky. 2021. "Deregulatory Science: Chemical Risk Analysis in Trump's EPA." *Social Studies of Science* 51 (1): 28–50.
Mitchell, Sandra D. 2009. *Unsimple Truths: Science, Complexity, and Policy*. University of Chicago Press.
Murphy, Michael Warren. 2020. "Notes toward an Anticolonial Environmental Sociology of Race." *Environmental Sociology* 7 (2): 1–12.
Moore, Jason W. 2011. "Transcending the Metabolic Rift." Journal of Peasant Studies 38 (1): 1–46.

Ollinaho, O.I. 2015. "Environmental Destruction as (Objectively) Uneventful and (Subjectively) Irrelevant." *Environmental Sociology* 2 (1): 53–63.
Oreskes, Naomi. 1999. *The Rejection of Continental Drift: Theory and Method in American Earth Science.* Oxford University Press.
Pearce, Trevor. 2010. "From 'Circumstances' to 'Environment': Herbert Spencer and the Origins of the Idea of Organism–Environment Interaction." *Studies in History and Philosophy of Science Part C: Studies in History and Philosophy of Biological and Biomedical Sciences* 41 (3): 241–52.
Pellow, David N. 2018. *What Is Critical Environmental Justice?* Polity.
Pickering, Andrew. 1995. *The Mangle of Practice: Time, Agency, and Science.* University of Chicago Press.
Pulido, Laura. 2016. "Flint, Environmental Racism, and Racial Capitalism." *Capitalism Nature Socialism* 27 (3): 1–16.
Rocheleau, Dianne. 1996. Gender and the Environment: A Feminist Political Ecology Perspective. Routledge.
Rosa, Eugene A., 2000. "Modern Theories of Society and the Environment: The Risk Society." In *Environment and Global Modernity,* edited by G. Spaargaren, A.P.J. Mol, and F. Buttel. Sage.
Rouse, Joseph. 2015. *Articulating the World: Conceptual Understanding and the Scientific Image.* University of Chicago Press.
Shtob, Daniel, and Jordan Fox Besek. 2021. "Environmental Precedent: Foregrounding the Environmental Consequences of Law in Sociology." *Sociological Forum* 36 (3): 712–34.
Soper, Kate. 1995. *What is Nature? Culture, Politics, and the Non-Human.* Wiley.
Sprenger, Florian, Erik Born, and Matthew Stoltz. 2023. "Surrounding and Surrounded: Toward a Conceptual History of Environment." *Critical Inquiry* 49 (3): 406–27.
Steinmetz, George. 2023. *The Colonial Origins of Modern Social Thought.* Princeton University Press.
Stuart, Diana. 2016. "Crossing the 'Great Divide' in Practice: Theoretical Approaches for Sociology in Interdisciplinary Environmental Research." *Environmental Sociology* 2 (2): 118–31.
Warde, Paul, Libby Robin, and Sverker Sörlin. 2018. *The Environment: A History of an Idea.* Johns Hopkins University Press.
White, Damian F., Alan P. Rudy, and Brian J. Gareau. 2016. *Environments, Natures and Social Theory: Toward a Critical Hybridity.* Macmillan.
Whyte, Kyle. 2018. "Settler Colonialism, Ecology, and Environmental Injustice." *Environment and Society.* 9:125–44.
Williams, Raymond. 1980. "Ideas of Nature." In *Problems of Materialism and Culture.* Verso.
York, Richard. 2017. "Why Petroleum Did Not Save the Whales." *Socius* 3.
York, Richard, and Brett Clark. 2010. "Critical Materialism: Science, Technology, and Environmental Sustainability." *Sociological Inquiry* 80 (3): 475–99.

2. Achieving a Transdisciplinary Turn in Environmental Sociology

Action-Oriented Theory

Matthew Houser

THE TROUBLE WITH MODERN ENVIRONMENTAL SCIENCE

I was recently fortunate enough to be invited to attend a local agribusiness' annual farmer workshop. Held in a Walmart-sized convention center and buffet, I sat alongside nearly four hundred dairy farmers, most of whom were Amish, an agrarian community that has, owing to its religious beliefs, intentionally secluded itself from modern technologies, practices, and thoughts (fig. 2.1). Before sending us off to a buffet lunch timed appropriately for a population that starts their workdays before dawn, a staff member of the company took the stage to address the group. He spoke about the importance of good communication and strong relationships. To illustrate the importance, he shared a story where communication went wrong. While visiting the Western United States, he witnessed a presentation where an ecologist affiliated with a local university spoke to a group of ranchers on how her research group would help them solve the issue of coyote predation on their cattle. She planned to tranquilize and then spay and neuter the coyotes, leading to an eventual population decline. The staff member then smiled, recounting how an "old-timer" in the back of the room raised his hand and, when called on, posed a critique of this solution: "Ma'am, I don't think you've understood our problem. The coyotes are trying to eat our cattle, not mate with them."

"Amish humor," as it is known, favors the inane, and it is not hyperbole to say quite a few of the audience members almost fell out of their seats laughing at this tale of misunderstanding and miscommunication. A farmer's child sitting beside me nearly aspirated his chocolate milk in glee.

While I certainly enjoyed the window into this reclusive community's humor, the deeper sentiment of the story was revealed to me as I discussed

FIGURE 2.1. Coat check at an Amish farmer meeting. Photo credit: Matthew Houser.

the event with several dairy producers over lunch. The farmers were not *just* laughing at the rancher's misunderstanding (though they found that amusing, too). They were laughing at the academic, who they felt proposed an "academic solution" for what was ultimately a practical problem. The ranchers in the story were losing cattle, risking their livelihoods *now*. They did not have the luxury of waiting for a gradual population decline over the long term. The ecologist, in an approach sometimes called "parachute science," *jumped* into a community's problem, did not seek the input of its members, and instead offered an ecologist's solution to the issue. This was seen as all too common to my lunchmates, who referenced several of the university-affiliated speakers earlier in the day—"that information was interesting and I can tell they are smart, but how does it help me?"

It has been widely observed that the perceived value of "scientific" knowledge for addressing contemporary socio-environmental problems has declined (Lave 2015). Explanations for this decline vary (e.g., Miroski 2011; Beck 1992). But central among these causes may be that—similar to the above ecologist—multiple disciplines in the environmental sciences too often miss the mark in terms of using their work to actually help communities address their problems.

While farmers are a specific example, their case is reflective of a larger challenge observed by other scholars in fields such as participatory-action research (Wylie et al. 2017). For many segments of the public, scientific approaches and "solutions" for socio-environmental challenges are often disconnected and ultimately unhelpful (or even harmful) (Wylie et al. 2014; see also chapter 6 of this volume). This is increasingly recognized by academic leaders and funding organizations, which are prioritizing efforts to better integrate scientists with communities and achieve action or change through more informed, connected research (Lang et al. 2012). And more specifically, the pressing challenge of addressing modern coupled systems or socio-ecological systems issues that involve social and environmental challenges—such as climate change, environmental quality degradation, and biodiversity losses—is leading to a reimagining of the practice of science (Wylie et al. 2017). No longer can science be disengaged, abstracted. In particular, the environmental sciences, including the environmental social sciences, must more effectively engage with diverse publics and, for this to happen, move toward a greater focus on producing change (Pennington et al. 2013; Minkler et al. 2008).

WHY WE NEED TRANSDISCIPLINARY ENVIRONMENTAL SOCIOLOGY

While the problem of disengaged science is not isolated to environmental sociology, environmental sociologists are ideally positioned to guide this transition within the sciences, but change must first start at home, in our own discipline.

To address the challenge of more actionable environmental science, a transdisciplinary approach is increasingly seen as a key means to enable more effective, actionable science related to socio-environmental crises (Lang et al. 2012). ***Inter*****disciplinary science** integrates scientific knowledge across disciplines, while ***trans*****disciplinary science** additionally incorporates societal perspectives (e.g., community members) into research in ways that both inform research design and motivate research to become more applied or actionable (Benard and De Cock-Buning 2014). In transdisciplinary research, community members are not just subjects, they are participants. This means that the social world must be better conceptualized and understood in emerging sustainability science, and that community's views and needs not only must be better incorporated into research, but that research must become better at leading to positive change in communities (Dodd et al. 2023).

Given environmental sociology's focus on understanding society-environment interactions, our discipline and its practitioners are extremely well positioned to guide the environmental sciences writ large toward transdisciplinary science. Although we are well-*positioned,* there is an opportunity to expand the degree to which environmental sociologists participate in transdisciplinary research.

That is not to say that environmental sociologists are not doing elements of transdisciplinary work. Yes, we work with communities to document their lifeworld. Yes, the distance between our subfield and that of other environment and society-focused fields is collapsing. And yes, some of our scholars collaborate with nonsociologists in team science. But if you have ever purchased an IKEA shelving unit, you already know that having the components of a thing is quite different from having the thing itself. I am arguing for a more systematic, intentional turn to transdisciplinarity in environmental sociology—that transdisciplinary, action scholarship should get a larger seat at the table.[1]

Achieving a more transdisciplinary environmental sociology requires numerous steps. Certainly, a key aspect will be building more effective teams of scientists that can address societal issues. This need and the barriers to interdisciplinary collaboration, along with other challenges across the environmental sciences, has been covered in much past work (see, e.g., Houser et al. 2021).

Here, I focus on a particular and less frequently discussed need within environmental sociology's transdisciplinary turn, focusing on the possibility of reimagining our subdiscipline's unique theory-focused research agenda. I will argue that our subdiscipline has a unique focus on theory (re)development, and while it is beneficial, it produces some barriers to authentic community engagement in transdisciplinary research. From this, I consider how our theory-focused research can evolve to become more action-oriented, or able to produce change, while maintaining our core character.

MAKING SOCIOLOGICAL THEORY APPLIED

Environmental Sociology's Unique Character

Environmental sociology has long been focused on macro-level theory building, and more specifically on showing the relevance of our subdiscipline to the core of sociology. Fredrick Buttel's (2002) wonderful summary of environmental sociology's unique positioning relative to natural resource sociology still articulates the majority of the subdiscipline's focus well. Buttel emphasized that environmental sociology strove to *understand* the social-environmental problem, not to address it. Indeed, the field's specific lack of focus on applied, solutions-oriented research is one of its defining points. In Buttel's view, environmental sociology (compared to natural resource sociology) was much more likely to be in conversation with the core discipline, an attempt to show the sociological relevance of social-environmental interactions through theory and meta-theory development (e.g., nationally, or globally focused).

While not all environmental sociology is theory focused, the testing and development of social theory is *at least* a major component of the work of an environmental sociologist, and it could be said that research that tests and reenvisions social theory is seen as the most prestigious contribution one could make to sociology and thus also to environmental sociology.

Writing for theory development and leveraging that theory in our work is valuable. Social theory helps us to understand and see elements of the social-ecological interaction process that we might miss; it enables us to connect our work to larger structural issues, binding otherwise disparate research topics together and revealing key barriers to socio-ecological progress writ large; it highlights the possibility for a better future, shedding light on the yet-unrealized potential of society. Practically, theory focused work is often publishable in our field's top journals. Theory-focused research is therefore key to tenure and promotion within most sociology departments.

The Problems with Theory

While environmental sociology's commitment to theory offers benefits, at the same time and to Buttel's point, theoretically focused work is generally distinct from applied-focused work. While theoretical research is intended to increase *understanding*, applied or "action research" is "deliberately intended to bring about social change" (Babbie 2013, 18). Focusing primarily on increasing understanding does present some barriers to authentic transdisciplinary engagement then. Theory-focused work is intended to shape the thinking and research activities of our academic peers, those who also read sociology journals. It is thus primarily intradisciplinary in its focus.

Moreover, and more significantly, authentic participation by community members in the research project is one of the most important but also most difficult aspects of transdisciplinary science (Altman 1995; Sommer 1977).

In terms of what will lead the public to participate, the nonacademic public is rarely concerned with what new advancement their problems help make to the field of critical sociology, or whether ecological modernization (Mol and Spaargaren 2000) or **treadmill of production** (Schnaiberg 1980) better describes the behavior of the local industry. People do not just want to better *understand* their problems; they want to see their lives improved through the process of science, especially the science that they participate in. Doing more applied "action research" is a key dimension of enabling the development of a transdisciplinary environmental sociology (Stokols 2006).

Environmental Sociology without Limits

Does our subdiscipline's uniquely theoretical take on society-environment interactions limit our capacity to become transdisciplinary? If we prioritize theory building for disciplinary development alone, then the answer is "yes." But this is by no means a predetermined destiny of our theory-focused work, and indeed, sociological theory for theory's sake has long been bemoaned.

Maybe most notably, Karl Marx (1845) pushed for the ability and need to create change in his *Theses on Feuerbach,* uttering the now famous (at least to sociologists) statement: "The philosophers have hitherto only interpreted the world in various ways, *The point, however, is to change it.*" While some have taken this to be a push for absolute activism, Marx, more accurately understood, is calling for a shift to studying "how [social] needs, interests, and powers shape and hold particular human conventions and *in which ways these conventions can be transformed*" (West 1991, 18; emphasis added). In short, he is calling for our sociological research into the barriers to social change, but also an explicitly action-oriented research agenda around changing it.

Marx's push for change-oriented theory and efforts is mimicked by the writings and actions of sociologists across the history of field. W. E. B. Du Bois, in his efforts to make his sociological theory on race, racism, and capitalism actionable, displays an unparalleled commitment to "action-oriented theory" (if you will). His writing, public speaking, political engagement, and his editorship of *The Crisis,* a widely read magazine in the twentieth century, are clear examples of his belief that sociologists should do more than theorize society. Du Bois, as Michael Burawoy (2021) articulates, "transcended the division between science and politics in both theory and practice. He gave public sociology [i.e., action-oriented sociology] pride of place in his vision of sociology, not antagonistic to professional, critical, and policy sociologies but as the driving force behind them" (28).

More recently, leaders at the core of sociology have expressed the need to increasingly connect sociology's critical theoretical insights and visions of a potential, more equitable, and more sustainable future with the public (Zussman and Misra 2007). For instance, Joe Feagin in his presidential address to the American Sociological Association (ASA) in 2001, expressed this need in the following statement: "Given the new century's serious challenges . . . sociologists need to think deeply and imaginatively about sustainable social futures and *to aid in building better human societies*" (1; emphasis added). Burawoy follows this tradition in his 2004 presidential address to ASA, and more recently in his 2021 book *Public Sociology: Between Utopia and Anti-Utopia*. In both, he makes a clear argument for the importance of more "public" or action-oriented sociology, seeing it as necessarily complementary to our more traditional, internally focused work. He gives particular attention to the role of critical sociology and the need to bring our theoretical insights about the problems and potential for society to a wider, transdisciplinary audience. In almost every case, these core leaders recognized the need to consider and change the environmental dimensions of our society as well.

These are only some of the sociologists who have acknowledged the need to do more than understand (see also chapter 4 of this volume). As their comments illustrate, sociology, and thus sociological theory, is not innately an intradisciplinary practice. Instead, at the core of our field, there is a long-standing recognition of the need to do more with our sociological insights—the need to apply it, to help change society, or in the work of an environmental sociologist, to change the status quo of social-ecological relationships.

Barriers to a Practical Turn in Our Theory-Driven Mission

This is not to say that everyone agrees that sociology, environmental sociology, and sociological theory should become more action-oriented, nor are we accounting for practical barriers to this change that many readers will encounter should they try to act toward this end.

To the former point, the issues of action-oriented science have long been debated (e.g., Nelkin 1977) and many have seen the shift to a more applied effort as detrimental to science generally and sociology specifically (Burawoy 2021). **Action-oriented science** is an approach to research that focuses on making a practical impact on communities and/or partners (Small and Uttal 2005). The environmental social sciences have broadly been argued to be necessarily and beneficially separate from the public, and specifically from any type of "advocacy" or action-oriented directions. The title of a 2001 article in *Human Dimensions of Wildlife* is quite illustrative of this opinion: "Science and Advocacy Are Different-and We Need to Keep Them That Way" (Neilson 2001). The thinking behind this position is that science, including social science, must remain untainted by subjective interests, and that as a neutral source of information, it is more capable of informing knowledge and societal direction (Ruggiero 2010).

Contemporary survey research suggests the American public is willing to accept scientific advocacy in the environmental realm with little impact on their trust in science (Kotcher et al. 2017). While this issue is more complex than this, as an environmental sociologist jointly appointed at a university *and* an environmental nonprofit, I have clearly made my choice in how I wish to proceed. With my bias noted, how a turn to "advocacy" or "applied science" in the work of environmental sociology will affect the discipline's legitimacy over time is unknowable and contextually dependent. We must consider this point as we turn toward or away from applying our discipline's theoretical visions.

Even if one wants to make a practical turn in environmental sociology's theory-driven world, they face numerous barriers built into the contemporary structure of academia. To be successful in tenure track positions, traditionally intradisciplinary achievements, especially publications in core journals and grants, are key. Public focused dimensions of our work—such as public writings, presentations, teaching classes, or even implementation metrics (such as acres of improved land, number of homes made more resilient, or economic/happiness improvements within a community)—generally have limited value in terms of professional advancement in a traditional sociology department. The contemporary professional reward structure limits the emergence of a transdisciplinary environmental sociology through limiting the emergence of more applied, translation version of our work (Houser et al. 2021).

Achieving a More Action-Oriented Theoretical Research Agenda

Given the difficulty of acquiring academic positions, and the often-limited job opportunities outside universities, it is unlikely that environmental sociologists in academic units will throw away professional security and advancement to take on more action-oriented work. How might we achieve applied-theoretical work, given the current reward structure of most traditional academic units?

Burawoy (2021) notes the potential for more focus on public writing, teaching, and community outreach. This is certainly a way forward and can be a means for the illuminating light of sociological theory to reveal the social dimensions of our intertwined socio-ecological issues today to community members. In many cases, I see that a new wave of social scientists and environmental sociologists appear to be more interested in these elements of action-oriented science. But more than the communication of our insights, transdisciplinary research that enables collaboratively determined social change is needed given the need to increase the pace and scale at which environmental and social justice practices are implemented by social actors. In other words, we must not just communicate, we must find a way for our research to directly help social actors achieve and become, while also advancing theory. We must achieve an action-oriented theoretical research program.

A subfield of study, called implementation science, may enable us to achieve an action-oriented theoretical research agenda (Hering 2018). **Implementation science**, from public health, was originally conceived as the "the scientific study of methods to promote the systematic uptake of research findings and other evidence-based practices into routine practice, and, hence, to improve the quality and effectiveness of health services and care" (Bauer et al. 2015, 6). Recent scholarship has emphasized that it conceptually lends itself to studying the process of making socio-ecological change (Hering 2018). Implementation science studies are structured by three main aspects (Hering 2018):

1. A *structured framework* (i.e., theory) is developed and applied to the process of implementing an evidence-based practice.
2. The socio-economic, political, cultural, and/or institutional *context* for implementation is considered.
3. Implementation is *monitored* to assess the intervention, the barriers, and the effectiveness of the practice in achieving its end.

As an example of how this structure plays out, Stoffels and colleagues (2021) examine, based on the *structure* of communication theory, the potential for message framing around water policy to reduce uncertainties and improve policymakers' decisions in New Zealand. However, their *monitoring* revealed that though they were able to reduce uncertainties, *contextual* factors constrained policymakers' capacity to prioritize science-based policy options in their policy decision-making.

As this illustrates, implementation science, more than simple communication of our work or even partnerships, opens the door for research projects that are informed by theory and test theory through actively piloting socio-ecological change initiatives (such as communication experiments, pilot policy programs, incentives that enable individuals/communities to try new practices, and norm experiments) with behavioral change as a primary outcome variable. It also considers a core element of environmental sociologists' interest—how contextual processes shape social activity. In other words, it pushes our transdisciplinary work to land on the ground with our community partners while also providing an avenue for traditional academic achievements, such as receiving grants and publishing theory-focused papers related to understanding this process. This is social research by doing, and doing that is being researched.

The practice of implementation science is not entirely novel to our subfield—far from it. But implementation science has not been systematically pursued within environmental sociology. This research topic and related methods (e.g., pilot intervention studies, especially) should become a more central research focus and can give language to environmental sociologists toward enabling them to acquire grant funding for action-oriented projects.

At the moment, the research field of implementation environmental science is growing, but it remains fairly non-theoretical when considering sociological perspectives. This work (and the communities it lands in) could benefit from sociological theory. Moreover, environmental sociology's core theory building mission may benefit through trialing specific change-based initiatives as well. There are opportunities in this work for sociological theory to shape the design of change initiatives and reflexively, for action research to be a means of testing and redeveloping sociological theories of social action and inaction.

The successful initiation of this action-oriented research field within environmental sociology depends in part on its ability to connect with and expand on our core theoretical debates and perspectives. While I believe the applicability is widespread, given space, I briefly outline one potential area of theoretical discussion where an implementation science approach could tie in.

How Do We Create Change Given Structural Barriers?

Much of our environmental sociology's core theoretical work has considered the role of social-structural conditions in (re)producing unsustainable and unjust socio-ecological relationships. These studies generally fall under the topical field of the political economy of the environment. Following in large part from Marxist thinking about the structure of modern capitalism, influential political economy texts in environmental sociology emphasize the degree to which structural conditions limit humans' capacity to achieve a sustainable world (Peet and Watts 2004). Allan Schnaiberg's (1980) concept of the **treadmill of production** has become the "ideal type" example of this theoretical tradition. In this work, Schnaiberg conceives a world where environmental improvements can never be achieved because production, which harms the environment both through resource extraction and pollution, will constantly expand due to the interrelated interests of labor, capital, and the state/politicians (Buttel 2004). This position is most famously countered by ecological modernization's vision of market-driven pro-environmental improvements over time (Mol and Spaargaren 2000). The two theories are often contrasted conceptually and through empirical analysis (York et al. 2010; York and Rosa 2003). Although this debate is assuredly not over, contemporary environmental sociology research generally emphasizes that the current profit-imperatives of our economic system put stark limits to achieving rapid, widespread, and effective pro-environmental change (among other desirable outcomes) (Adua et al. 2021; Jorgenson and Clark 2012).

Recent work has begun to consider how to move beyond focusing on the limits of our system. Eric Olin Wright's *Envisioning Real Utopias* (2020) emphasizes the potential to create change from within, toward transforming the capitalist structure to achieve a better, juster, world. He argues that this change starts with "real utopias," those spaces that exist contrary (to some extent) to capitalism's structure, but are nonetheless within the system. Capitalism developed on the fringes of feudalism. What comes after capitalism will develop on the fringes of our current socio-economic structure.

Wright gives us the conceptual lens to think beyond the limited potential for pro-environmental change within capitalism. Indeed, Wright's perspective suggests that focusing only on the limits to pro-environmental change is itself a distraction from achieving change. His work asserts strongly that it is possible and that we, as (environmental) sociologists, must continue to explore *how* social change occurs. Wright championed efforts to create change from within, and specifically noted the potential of interventions such as universal basic income.

This is a call to move beyond conceptualizing the barriers and begin to work toward the change. It is also a theoretical finger pointing to the potential of implementation science studies. Recent work has already begun applying Wright's framework to examine environmental social movements (Stuart et al. 2020). The structure of implementation science allows us to move beyond examining how change is happening toward testing ways to create change. For an implementation environmental sociology subfield, this suggests the following broad structuring question: *What interventions can effectively promote the development of positive socio-ecological change, given structural barriers?* This question can certainly be applied to individual-level behaviors, but also is equally applicable to other scales of social change, including communities, social movements, or policy.

An "implementation environmental sociology" research agenda asks us to pursue questions in line with this notion through piloting the impact of innovative social-ecological interventions that are theorized to enable a study population to think, act, or be more sustainable, resilient, and equitable. For instance, we can examine to what extent and how interventions such as universal basic income or direct "incentive" payments can enable communities or individuals to pursue socio-ecological behavior change; whether risk reducing strategies, such as innovative insurance mechanisms, can help businesses trial pro-environmental and/or social welfare initiatives despite capital's structuring impact; and how access to certain social movement resources encourage the spread of these movements.

These are only potential ways that an implementation science research agenda can empirically engage environmental sociological theory. Our success in addressing this and further questions can practically help communities, while also enabling us to build social-change theory for further testing. Our failures can shed new light on traditional sociology debates, including how material or ideological factors constrain meaningful socio-ecological progress.

CONCLUSION

To address today's environmental challenges, the environmental sciences writ large must become increasingly committed to working with communities toward achieving collaboratively defined goals and real-world change. Environmental sociologists can, through their participation in interdisciplinary projects, help to lead this shift. But first, we must give elements of transdisciplinary research a larger seat at the table within our subdiscipline. While many steps must be considered, an implementation science research agenda may allow us to capitalize on the unique potential of our theory-focused research while overcoming the barriers to action-oriented outcomes this theory-focus presents. With a focus on action and on-the-ground impacts, an "implementation environmental sociology" field allows us to explore to what extent various intervention strategies produce socio-environmental benefits given the structural barriers of modern capitalism—and to what degree we can overcome these constraints through various interventions. Challenges to this shift certainly exist—should sociologists engage in advocacy-type science? Can we actually pursue this type of research given institutional reward systems? And, of course, can we get grants to support this work? I think the answers broadly are "yes," but I make no claims that this will be easy—only that it is worth it.

Ultimately, environmental sociology and the communities we work with would benefit greatly by driving this research topic further. Connecting our subfield to transdisciplinary collaborations, where community members can help define what goals they wish to achieve, could dramatically increase the degree to which sociological insights about society's potential become realized.

NOTE

1. Notably, environmental sociologists often practice transdisciplinary scholarship. However, this practice is rarely taught in environmental sociology programs, nor is it widely considered highly valued within our traditional rewards and cultural esteem systems in the modern environmental sociology department. Indeed, if one looks outside an environmental sociology department or journal, one will see many environmental sociologists practicing transdisciplinary, action-oriented research. What I push for is the acceptance of this work at the core of our field, rather than on the periphery.

REFERENCES

Adua, Lazarus, Brett Clark, and Richard York. 2021. "The Ineffectiveness of Efficiency: The Paradoxical Effects of State Policy on Energy Consumption in the United States." *Energy Research & Social Science* 71:101806.

Altman, David. G. 1995. "Sustaining Interventions in Community Systems: On the Relationship between Researchers and Communities." *Health Psychology* 14 (6): 526–36.

Babbie, Earl R. 2020. *The Practice of Social Research*. Cengage Learning.

Bauer, Mark S., Laura Damschroder, Hildi Hagedorn, Jeffrey Smith, and Amy M. Kilbourne. 2015. "An Introduction to Implementation Science for the Non-Specialist." *BMC Psychology* 3 (32).

Beck, Ulrich. 1992. *Risk Society: Towards a New Modernity*. Sage.

Benard, Marianne and Tjard De Cock-Buning. 2014. "Moving from Monodisciplinarity Towards Transdisciplinarity: Insights into the Barriers and Facilitators that Scientists Faced." *Science and Public Policy* 41 (6): 720–33.

Burawoy, Michael. 2021. *Public Sociology*. Polity.

Buttel, Frederick H. 2002. "Environmental Sociology and the Sociology of Natural Resources: Institutional Histories and Intellectual Legacies." *Society & Natural Resources* 15 (3): 205–11.

Buttel, Frederick H. 2004. "The Treadmill of Production: An Appreciation, Assessment, and Agenda for Research." *Organization & Environment* 17 (3): 323–36.

Dodd, Warren, Sara Wyngaarden, Sally Humphries, Esmeralda Lobo Tosta, Veronica Zelaya Portillo, and Paola Orellana. 2023. "How Long-Term Emancipatory Programming Facilitates Participatory Evaluation: Building a Methodology of Participation through Research with Youth in Honduras." *Action Research* 22 (3): 243–61.

Feagin, Joe R. 2001. "Social Justice and Sociology: Agendas for the Twenty-first Century: Presidential Address." *American Sociological Review* 66 (1): 1–20.

Hering, Janet. 2018. "Implementation science for the environment." *Environmental science & technology* 52 (10): 5555–60.

Houser, Matthew, Abigail Sullivan, Tara Smiley, Ranjan Muthukrishnan, Elizabeth Grennan Browning, Adam Fudickar, Pascal Title, Jason Bertman, and Maria Whiteman. 2021. "What Fosters the Success of a Transdisciplinary Environmental Research Institute? Reflections from an Interdisciplinary Research Cohort." *Elementa Science of the Anthropocene* 9 (1): 00132.

Jorgenson, Andrew K., and Brett Clark. 2012. "Are the Economy and the Environment Decoupling? A Comparative International Study, 1960–2005." *American Journal of Sociology* 118 (1): 1–44.

John E. Kotcher, Teresa A. Myers, Emily K. Vraga, Neil Stenhouse, and Edward W. Maibach. 2017. "Does Engagement in Advocacy Hurt the Credibility of Scientists? Results from a Randomized National Survey Experiment." *Environmental Communication* 11 (3): 415–29.

Lang, Daniel J., Arnim Wiek, Matthias Bergmann, Michael Stauffacher, Pim Martens, Peter Moll, Mark Swilling, and Christopher J. Thomas. 2012. "Transdisciplinary Research in Sustainability Science: Practice, Principles, and Challenges." *Sustainability science* 7:25–43.

Lave, Rebecca. 2015. "The Future of Environmental Expertise." *Annals of the Association of American Geographers* 105 (2): 244–52.

Marx, Karl, and Friedrich Engels. 1968. *The German Ideology*. Progress.

Minkler, Meredith, and Nina Wallerstein, eds. 2011. *Community-Based Participatory Research for Health: From Process to Outcomes*. John Wiley & Sons.

Mirowski, Philip. 2011. *Science-Mart: Privatizing American Science*. Harvard University Press.

Mol, Arthur P. J., and Gert Spaargaren. 2000. "Ecological Modernisation Theory in Debate: A Review." *Environmental Politics* 9 (1): 17–49.

Nelkin, Dorothy. 1977. "Scientists and Professional Responsibility: The Experience of American Ecologists." *Social Studies of Science* 7 (1): 75–95.

Peet, Richard, and Michael Watts, eds. 2004. *Liberation Ecologies: Environment, Development, Social Movements*. Routledge.

Pennington, Deana D., Gary L. Simpson, Marjorie S. McConnell, Jeanne M. Fair, and Robert J. Baker. 2013. "Transdisciplinary Research, Transformative Learning, and Transformative Science." *BioScience* 63 (7): 564–73.

Ruggiero, Leonard F. 2010. "Scientific Independence and Credibility in Sociopolitical Processes." *Journal of Wildlife Management* 74 (6): 1179–82.

Schnaiberg, Allan. 1980. *The Environment: From Surplus to Scarcity*. Oxford University Press

Small, Stephen A., and Lynet Uttal. 2005. "Action-Oriented Research: Strategies for Engaged Scholarship." *Journal of Marriage and Family* 67 (4): 936–48.

Sommer, Robert. 1977. "Action Research." In *Perspectives on Environment and Behavior: Theory, Research, and Applications*, edited by D. Stokols. Plenum Press.

Stoffels, Rick J., Paul A. Franklin, Stephen R. Fragaszy, Doug J. Booker, Joanne E. Clapcott, Ton H. Snelder, Annika Wagenhoff, and Chris W. Hickey. 2021. "Multiple Framings of Uncertainty Shape Adoption of Reference States During Reform of Water Policy." *Environmental Science & Policy* 124:496–505.

Stokols, Daniel. 2006. "Toward a Science of Transdisciplinary Action Research." *American Journal of Community Psychology* 38:63–77.

Stuart, Diana, Ryan Gunderson, and Brian Petersen. 2020. "The Climate Crisis as a Catalyst for Emancipatory Transformation: An Examination of the Possible." *International Sociology* 35 (4): 433–56.

Wylie, Sara Ann, Kirk Jalbert, Shannon Dosemagen, and Matt Ratto. 2014. "Institutions for Civic Technoscience: How Critical Making Is Transforming Environmental Research." *The Information Society* 30 (2): 116–26.

Wylie, Sara, Nick Shapiro, and Max Liboiron. 2017. "Making and Doing Politics through Grassroots Scientific Research on the Energy and Petrochemical Industries." *Engaging Science, Technology, and Society* 3:393–425.

West, Cornell. 1991. *Ethical Dimensions of Marxist Thought*. Monthly Press Review.

Wright, Erik Olin. 2020. *Envisioning Real Utopias*. Verso Books.

York, Richard, Eugene Rosa, and Thomas Dietz. 2010. "Ecological Modernization Theory: Theoretical and Empirical Challenges." In *The international Handbook of Environmental Sociology*, edited by M. Redclift and G. Woodgate. 2nd ed. Edward Elgar.

Zussman Robert, and Joya Misra. 2007. Introduction to *Public Sociology: Fifteen Eminent Sociologists Debate Politics and the Profession in the Twenty-First Century*, edited by D. Clawson. University of California Press.

3. Animalizing Environmental Sociology

Cameron T. Whitley and Abraham Vanselow

Humans have long relied on nonhuman animals for food, companionship, labor, transportation, and safety, all of which have contributed to the development of society (Kalof and Whitley 2021). Despite their undeniable integration in social systems and society at large, animals have tended to be "defined as the concern of biology rather than sociology" (Carter and Charles 2018, 85). But, when animals are restricted to the world of the natural sciences, we miss the opportunity to study human-animal and social-environmental phenomena—for example, the impact of animal feces on community water quality (Mori et al. 2023), food webs and declines in pollinator populations (Hale, Valdovinos, and Martinez 2020), and shared suffering among species in environmentally depleted environments (see, e.g., Whitley 2019).

In this chapter we argue for the further animalizing of environmental sociology by pointing out that, while environmental sociology seems receptive to greater integration of nonhuman species, little integration has been done since the inception of the subdiscipline. As Richard Twine (2020) attests, animalizing environmental sociology is critical to understanding the complexities of environmental problems. For instance, we cannot fully understand greenhouse gas emissions without understanding factory farming, carnism, or animal exploitation. Similarly, we cannot detail an analysis of the extractive energy industry without understanding how animals were integral to energy development or how the air and water pollution hydraulic fracturing creates results in shared suffering among humans and nonhumans (Whitley 2017, 2019). Learning about sociological animal studies also enhances concern not just for animals but for the environment and other humans as well (Whitley et al. 2024).

Broadly, **animalizing** is the becoming or embodying of nonhuman animals. In the humanities, this equates to conceptualizing how an animal thinks, feels, moves, and engages the world around them. In the social sciences, animalizing is not about the physical embodiment of an animal but the expansion of the subdiscipline to center nonhumans as those who create and share in "mutually experienced environmental outcomes" (Stuart and Gunderson 2020). To do this, we need a radical rethinking of ourselves as students and researchers in a context that includes nonhuman animals. To do so, we look to queer theory to move beyond the dichotomy of dualism or monism, into a polysemic scholarly approach that goes beyond a discussion of value-free research and limited engagement in activism, to the recognition that (1) our engagement with environmental issues (and animals broadly) is likely driven by a desire to support or advocate for these communities; (2) social science is inherently human centered; (3) nonhuman animals occupy multiple realities of human existence (companion, food, entertainment, sentinel, etc.); (4) the inclusion or exclusion of nonhuman animals has impacts on the material realities of nonhuman animal lives; (5) nonhuman animals are distinct from other sentient participants and subjects because they have few legal protections; and (6), that the inclusion of nonhuman animals is important because we share mutually experienced environmental outcomes.

HISTORICAL INCLUSION AND EXCLUSION OF NONHUMAN ANIMALS

The Historical Exclusion of Nonhuman Animals from Sociology

Sociology has historically excluded nonhuman animals. Some assign blame to George Herbert Mead and his views that "symbolic interaction could only take place when interactants possess a sense of self—and language to communicate about it," something that he believed only humans possessed (Taylor and Sutton 2018, 468). But, looking back further, we can see elements of René Descartes's ideas about animals in Mead's theories—specifically, Descartes's belief that animals lack the ability to think and speak. These ideas can be traced back further to his "Scala Naturae [or] Great Chain of Being," which is "a hierarchical paradigm situating humans, specifically property-owning men, as rulers over the material realm—a realm where other animals permanently remain separate and beneath mankind" (Ross 2017, 5; Kalof and Fitzgerald 2007). What we find here is that sociology has been heavily influenced by colonialism, capitalism, anthropocentrism, and human exceptionalism. In her explanation of why animals are so often missing in sociology, Rhonda Wilkie (2015) blames the "anthropocentric culture that generally legitimates, objectifies, and normalizes the use of animals for human purposes" that many scholars have been raised in (324). Within environmental sociology, this objectification places animals in the same category as natural resources like oil or grains, when animals are actually sentient beings.

While limited in scope, some classical social theorists did engage with the natural world, occasionally incorporating animals into their analyses. For example, in the *Philosophic and Economic Manuscripts of 1844*, Karl Marx (1988) conceived of humans as a part of the natural world, stating that "Man *lives* on nature, means that nature is his *body*, with which he must remain in continuous intercourse if he is not to die. That man's physical and spiritual life is linked to nature means simply that nature is linked to itself, for man is a part of nature" (76). Additionally, early Frankfurt School theorists, inspired by Marx, saw the exploitation of animals and the natural world as "intimately linked to the domination of human beings, especially of women and racial and ethnic minorities" (Gunderson 2014a, 286). Since then, as we see from the increase in sociological papers with animals as the central topic, there has been a shift toward greater inclusivity of sociological animal studies, although there is still much work to be done, especially in the move toward an emancipatory animal sociology (see, e.g., Taylor and Sutton 2018), which calls for research to challenge the oppression of animals in society.

Nonhuman Animals as Historically Overlooked in Environmental Sociology

Environmental sociology developed as a subdiscipline from debates challenging anthropocentrism (Dunlap 1980; Dunlap and Catton Jr 1979; Dunlap and Van Liere 1978). Yet, while it was established to include the natural environment as a factor in social analysis, with the premise being that humans both impact and are impacted by the environment, there was little distinction made between nonsentient flora (plant) and fauna (animal). All were treated as natural or broad environmental features. As an example, in a detailed assessment of the inclusion of animals in published sociological climate change work over the past decade, Twine (2020) found that nonhuman animals remained nearly absent from the discourse. Specifically, most leading sociological climate change textbooks contain only fleeting mentions of nonhuman animals. These mentions include suggesting that factory farming is contributing to greenhouse gas emissions and that climate change is contributing to ecological declines. A similar trend emerges in the review of sociological climate change in peer-reviewed journal articles over the past decade. Twine (2020) found that none of the articles specifically about climate change featured in various journals "included a significant consideration of nonhuman animals or human/animal relations in emergencies, impacts, or mitigations of climate change" (122). The author defined a significant consideration as the inclusion of variables or data related to nonhuman animals in modeling or analysis. Only two articles briefly mentioned emissions from animal agriculture. Nine of the articles briefly discussed how humans and animals are both impacted by climate change. A further analysis of sociological climate change articles in the *Environmental Sociology* journal revealed a similar trend, with only one of the thirty-eight identified having any significant focus on animals. While climate change is only one area of environmental sociology, its connection to animals is inherent, while the connection between animals and other environmental issues may be less apparent, making it highly likely that they would indicate a similar trend. This is problematic given the research asserting that animals are sentient and engage in action, agency, and resistance and that some of their actions challenge human perception and may induce change (Carter and Charles 2013). Looking at value orientations, valuing the environment is distinct from valuing animals, where some people may value animal life, but not care about the environment or even substantially about animal life (Dietz, Allen, and McCright 2017).

Contemporary Considerations of Animals in Environmental Sociology

Although the social sciences generally and environmental sociology specifically have become increasingly open to animals and interdisciplinary animal studies courses, Twine's (2020) analysis demonstrates that being *open* to including animals does not necessarily translate into their incorporation in research and teaching. It is likely that few environmental sociologists were taught how to consider and include animals in their research and teaching (Grauerholz et al. 2024). In the social sciences, research incorporating animals has largely assessed the meanings humans make of animals, rather than their material reality (York and Longo 2017). This distinction is critically important. For example, continuing with the climate change case, research has assessed how animals are used in climate change narratives (Whitley and Kalof 2014), but this work, while important, does not speak to the material reality that animals are both contributing to and affected by climate change. Importantly, we do not intend to suggest that examining the social construction of animals is invalid or unnecessary; rather, we endorse additional ways of including animals in sociological analyses by grounding them in realism, something that many environmental and sociological animal studies scholars agree with (Kalof and Whitley 2021; Stuart and Gunderson 2020; York and Longo 2017). Diana Stuart and Ryan Gunderson (2020) make this argument, writing that scholars need to "expand beyond cultural studies of animals (e.g., how humans find meaning from animals) to examine humans and animals in terms of ecological relationships and environmental outcomes" (69). By focusing on animals' material reality, we can begin to examine how human-animal interactions "create mutually experienced environmental outcomes" (Stuart and Gunderson 2020, 69). Examples of these outcomes include the following: how companion animal waste can damage urban greenery and increase levels of airborne bacteria and compromise waterways for all living species (Bowers et al. 2011; Romo et al. 2019); how animal agriculture is a leading cause of greenhouse gas emissions, as well as a leading contributor to human and animal suffering (Economic Research Service 2022; Herrero et al. 2011); and how urban human-wildlife conflict is increasing owing to habitat encroachment and forced species migration (Woody 2021). While these are only a few examples, they highlight the importance of looking beyond categorizations and meaning-making into lived experiences across species.

ACTIVISM AS A PILLAR OF ENVIRONMENTAL SOCIOLOGY

Mixing Science and Activism for the Environmental Sociologist

Normative debates should be at the center of environmental and animal sociology as we consider the impacts of environmental issues; this, however, can be controversial (Whitley and Cherry 2023). As chapter 4 of this volume argues, sociology as a science is premised on conducting value-free research that can be used to address social ills. Although this focus may have increased the legitimacy of the field for a time, it can hinder progress on important issues by muting essential commentary from scholars. Andrew Abbott (2018) contends that this creates a dualistic approach for sociological scholars, where they attempt to divorce themselves from the value orientation that led them to conduct research in the first place. In practice, the topics scholars study are often connected to their core values, but those core values are not openly disclosed or discussed. This leads scholars to pursue topics that are important to them, while simultaneously attempting to neutralize their interest to present a value-free narrative.

FIGURE 3.1. Bengal tiger. Image by Tim Flach.

This effort reflects the fact that scholars have long argued that remaining value-neutral is the best approach for research objectivity. However, some subdisciplines, like environmental sociology and sociological animal studies, must challenge this contention because the very topics of study depend on ethical dilemmas and value determinations (Dietz, Shwom, and Whitley 2020). Importantly, this engagement with ethical concerns is not new but has deep historical roots. Gunderson (2014b, 2018), for instance, highlights environmental sociology's long engagement with ecological ethical decision-making by discussing normative connections that date back to the Frankfurt School and their concerns about the exploitation of animals.

For environmental sociologists (and sociological animal studies scholars), the question is whether it is enough to simply conduct and present research findings or whether there is a point at which one can and should use their platform to advocate for change. The answer to this question varies, but it increasingly includes value-acknowledgment and advocacy. For many within sociological animal studies and environmental sociology, there is little separation between the self as activist and the self as scholar. For instance, sociological animal studies and environmental sociology scholar Cameron T. Whitley (the lead author of this chapter) has used his research to advocate for change. He became a sociologist to use his expertise to help the environment and animals. He currently works with conservation photographers like Tim Flach, as well as with zoo and aquarium professionals, and policy scholars to identify how to elicit empathy for animals and promote pro-environmental behavior change. By working with conservation photographers, he has shown that animal portraiture (see figs 3.1 and 3.2) works best to elicit an emotional response and encourage behavior

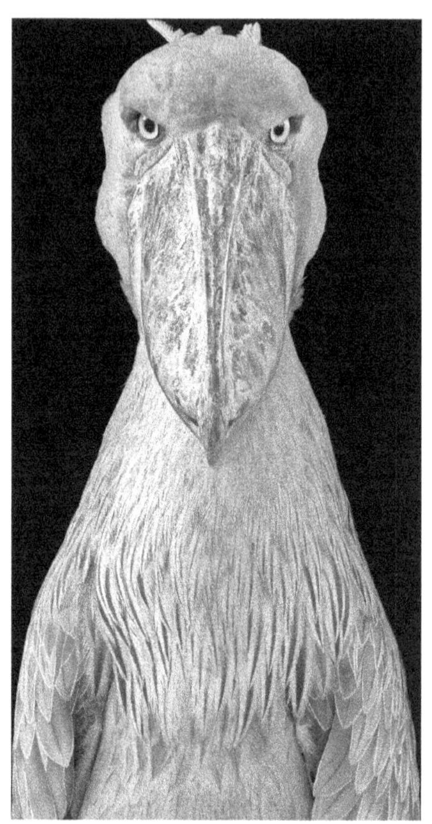

FIGURE 3.2. Shoebill stork. Image by Tim Flach.

change (Whitley, Kalof, and Flach 2021). He uses his research to write commentaries, support marketing campaigns, connect with policy professionals, volunteer for conservation organizations, and protest harm against the environment and animals. His research is driven by activism, and his activism is driven by his research.

Similarly, environmental sociology scholars like Andrew Jorgenson have advocated for environmental protections. Although he does not consider himself an activist, he explains that he was driven to this work because of concern for the planet and its inhabitants (Jorgenson 2018). Nik Taylor (2012) argues that the expanse of sociological animal studies has created room for more scholars to feel empowered in doing research that is activist-oriented. Scholars within and outside sociology are increasingly engaging in a variety of activist pursuits, such as participating in global marches, testifying before congress, demonstrating for the support of science in addressing climate change, and even developing scholar-activist identities (Frank 2017; Harmon and Fountain 2017; Kahn 2016; MacKendrick 2017). These approaches seem to be opposed to the dualism that is often expected of social scientists and so, as an alternative, Andrew Abbott (2018) argues for monism, suggesting that we cannot divorce ourselves from our values or the reasons we pursue research topics. This means that there is no value-free sociology, as the discipline was founded on normative pursuits with the goal of changing or advancing society. He suggests that we are one with our identities and leanings even when conducting research, but that scholar-activists can still conduct sound research. These two paradigms reflect an all-or-nothing mentality, which may not reflect the realities of most academics. Perhaps, when considering global social problems that demand expertise and broad social networks, scholars do not need to select between binary categories to organize their research and personal values as divorced or connected, but they can consider the personal, political, and social costs involved in these endeavors.

QUEERING THE BOUNDARY BETWEEN RESEARCH AND ACTIVISM

From a queer theory perspective, there are three leading challenges to Abbott's monism. First, Abbott (2018) suggests that the monist and dualist approaches are opposites that do not coexist or overlap. As Nik Taylor (2012) and Andrew Jorgenson (2018) demonstrate, labeling and articulating the ethical concerns driving inquiry is situational and often politically driven. While some situations and studies warrant this reflection, others do not. In fact, researchers may use their findings in different ways to appeal to distinct audiences. In this way, adhering strictly to either a value-neutral or value-laden approach may be disadvantageous for addressing environmental and animal issues. Second, Abbott's (2018) assertion that we cannot divorce ourselves from our values or the reasons we pursue research topics does not recognize that some may not be in a position or institution that allows for advocacy without consequence. In this way, being a public sociologist, scholarly activist, or research advocate is a privilege and some may need to completely or strategically "divorce" themselves from their activist leanings. Third, Abbot's (2018) depiction focuses on the researcher exclusively, without the recognition that other sentient beings (humans, animals, and nature broadly) are participants in the research process and that the welfare and needs of these other beings may dictate the process used. As a result, a focus on dualism or monism limits environmental sociology from being animalized. To be animalized, environmental sociology would need to recognize that nonhuman animals serve multiple roles in our lives—some exploitative, some coerced, and some consensual—but that they largely lack the ability to exit exploitation and that we are continuously creating and sharing in "mutually experienced environmental outcomes" (Stuart and Gunderson 2020). Nonhuman animals are not subjects that we can dismiss or deny as being fully engaged and embodied partners in our lived experiences or research.

Animalizing Environmental Sociology with Queer Theory

One way to get at animalizing environmental sociology is by using queer theory. Environmental sociologists and sociological animal studies scholars have called for increased engagement with queer theory to assess how gender, sexuality, and binary thinking influences how we think about and engage environmental and animal-related issues (Bowers and Whitley 2024; Dietz et al. 2020; Whitley and Bowers 2023). **Queering** comes out of queer theory, which was once centered on gender and sexuality (Halperin 2003; Jagose 1996; Morland and Willox 2017; Sullivan 2003), but has been expanded to consider how heteronormativity and systems of power influence systems of oppression that are situated in binary thinking. Queering rejects simplistic binaries and boundaries in favor of expansive categories, connections, and challenges to oppressive systems. Relatedly, scholars have asserted that environmentalism and concern for environmental issues have largely operated on a heteronormative (hetero-ecological) platform (Mortimer-Sandilands and Erickson 2010; Seymour 2013). Queer ecology seeks to question binary constructions of ecology—for example, in things like reproductive decision-making and how we treat the choice to have nonhuman animals over human children (Sbicca 2012). The realities of the climate crisis, biodiversity loss, animal exploitation, and other environmental ills highlight the importance of the activist-scholar who is willing and able to use their research to advocate for change. The reality, though, is that to engage in this work, we must move beyond conceptions of research and activism as siloed activities, as either activist-scholars or scholars simply. Although Abbott (2018) makes significant progress towards this goal by arguing that there is no value-free sociology because we cannot divorce ourselves from our values, we argue that we must go further in our conception of both practices and their interaction in the world by recognizing and acting on the ethical imperative to use our research for activism as necessary and when it is safe to do so.

Research is not just about our ethical perceptions of self; it is also about the ethical dimensions of our work with others. We contend that to animalize environmental sociology, we must adopt a perspective that recognizes the links between the researcher and the nonhuman animal subject in the push to address environmental ills. Just as animal sentinels are key to our understanding of environmental toxins (Mattes and Whitley 2021; Whitley 2017, 2019), nonhuman animals are also at the forefront of environmental crises, serving not just as sentinels for humans but also being independently affected. It is our job to join them in their struggle by including their stories and by using our research to advocate for their lives. To understand nonhuman animals not just as participants in our research but as individuals whose limited rights and legal protections require our advocacy for survival requires a reimagining of the research-advocacy process that we term *polysemic* engagement. Without moving beyond dualism and monism, we cannot animalize environmental sociology.

THE SIX PILLARS OF POLYSEMIC SCHOLARLY ENGAGEMENT

The word "polysemic" suggests several meanings. In the context of sociological animal and environmental research and advocacy, this starts with the recognition, consistent with monism, that we come to research because of our values and interests. While we engage in systematic scientific methods and processes designed to support replicability, our work may have multiple meanings and is not value-free (as dualism suggests). For many, systematic research has meaning both within an academic context and within an advocacy or policy context because our pursuit of environmental and animal related research is driven by a desire to support and protect these communities. Within the context of sociological animal studies and environmental sociology, it is the recognition that these interests and approaches are tied to critical environmental issues that impact sentient subjects, participants, and communities with few, if any, legal rights. **Polysemic scholarly engagement** (PSE) is not just about a stance on subjectivity and research semantics; it is also about the impact of research on animal exploitation or liberation. We outline six pillars of PSE that, if incorporated into our conception of research, could promote the animalizing of environmental sociology. PSE is the recognition that:

1. Our perceptions and perspectives remain human-centered (human-centric)
2. Nonhuman animals occupy an infinite number of social meanings and material realities in society (occupy infinite realities).

3. The inclusion or exclusion of animals in sociological research has consequential impacts on the material reality of animals (inclusion or exclusion is consequential).
4. As scholars or students, concern and value for environmental and nonhuman animal issues led us to engagement with nonhuman animals and environmental issues (values drove us to nonhuman animals).
5. The nonhuman animal subject is distinct from other humans and environmental processes being investigated in terms of legal recognition and protection (nonhuman animals have limited protections).
6. We engage in "mutually experienced environmental outcomes" with nonhuman animals owing to environmental ills (mutually experienced environmental outcomes).

Scholars and activists have long called for the emancipation and liberation of animals. Animalizing our approaches gets us closer to these realities.

ANIMAL EMANCIPATION AND LIBERATION AS POLYSEMIC ENGAGEMENT

Focusing on PSE as scholars and students helps us get closer to ending systematic oppression across species. The terms "emancipation" and "liberation" are often used interchangeably in the context of nonhuman animal lives; however, they have slightly different meanings. **Liberation** is the act of setting someone (human or nonhuman) free from imprisonment, confinement, or slavery. **Emancipation** involves setting someone free from legal, social, or political restrictions. Liberation is the physical manifestation of emancipation. In line with Yasmin Koop-Monterio's (2021) suggestion for studying animals as an oppressed group, Nik Taylor and Zoei Sutton (2018) call for an "emancipatory animal sociology" to recognize that human-centric systems continuously dominate nonhumans even in environments that are designed to support them such as rescues or sanctuaries. In looking back at environmental sociology, the lack of engagement with nonhuman animals demonstrates this sentiment. Much of the work on animal emancipation and liberation is embedded in **critical animal studies** (CAS) (McCance 2012), which is an interdisciplinary field that emerged out of the animal liberation and animal rights movements (Nocella et al. 2014) with ties to Peter Singer's *Animal Liberation* (2012). The discipline centers on activism and makes little distinction between scholarship and advocacy.

Total liberation as an applied theoretical concept emerged in the early 2000s to recognize and unite distinct nonhuman animal, human, and environmental social justice movements seeking to address shared and overlapping oppressive systems (Best 2014; Pellow 2014). It is considered a commitment rather than a blueprint for reform. Animalizing sociology broadly should be built on considering ways to promote the total liberation of nonhuman animals. Scholars assert that a "total liberation ecology" may be necessary to center the exploitation of the nonhuman world (Cochrane 2012). Total liberation is best used to assert that oppressions, even across species, are inherently interconnected and must be recognized as such for liberation of all to occur. The idea of total liberation also recognizes that human and nonhuman animals can be coconspirators in liberating each other (Allen and von Essen 2018). As Lori Gruen (2007) asserts, "Women, people of color, queers, non-human animals are all thought to be lower in the hierarchy than white, heterosexual, able-bodied men . . . Oppression of any of these groups is thus linked, and if one is opposed to sexism, racism, and heterosexism, etc. she should also oppose speciesism" (336). While the total liberation concept and movement was created to be inclusive, it has faced criticism because it considers animal and human liberation as morally equivalent, with equal consideration based on an acknowledgement of sentience. Yet human liberation is, by necessity, a precondition for animal liberation (Fotopoulos and Sargis 2006). Polysemic engagement allows us to recognize that the liberation of any community of species enhances the potential liberation of others.

CONCLUSION AND MOVING FORWARD

Regardless of activist orientation, it is increasingly clear that excluding nonhuman animals from environmental sociology, and sociology broadly, has negative implications. At the intersection of sociological animal studies and environmental sociology, sociologist Leslie Irvine provides one of the starkest examples of this reality. Investigating human and nonhuman animal loss during Hurricane Katrina in 2005, Irvine (2009) documents how restrictions on evacuating with animals led to a greater loss of human life, as people often refused to evacuate without their companion animals. Ultimately, this reality translated into the introduction of the Pets Evacuation and Transportation Standards (PETS) Act of 2006, which authorizes the Federal Emergency Management Agency (FEMA) to provide shelter and rescue for companion animals during major disasters and emergencies. By pushing beyond disciplinary norms and including animals in her analysis, Irvine was able to uncover a central problem for human evacuation; in so doing, Irvine also highlighted the distinct problem companion animals that were left behind faced as they were lost or perished.

Sociology generally, and environmental sociology specifically, needs to follow Irvine and other scholars' lead in incorporating animal concerns into their work, not simply in ways that reflect nonhuman animals impacts on humans, but with the recognition that human choices create sustained impacts for animal life that nonhumans have virtually no power to defend themselves against. Doing this helps to demonstrate the inextricable links between the human and nonhuman worlds, the degree to which our fates are interdependent, and the absolute need that the world has for scholars to make meaning from their work by pursuing activism. Given that the climate crisis will increasingly become a central focus within environmental sociology and across all social science fields, ensuring the well-being of animals—companion, liminal, wild, or otherwise—will require a coordinated fight against climate change (Lacetera 2019). Without attempts to reduce greenhouse gas (GHG) emissions, any efforts focused on improving or maintaining the material conditions of animals (human and nonhuman) will be futile. The combined efforts of environmental sociology and sociological animal studies have the potential to transform sociology broadly to better include the natural environment and nonhuman inhabitants in the analysis of social life and to address climate crisis mitigation and adaptation efforts. While philosophical differences among some environmental and animal studies scholars are likely to prevail, there is no denying that our commodification of animals has implications for the climate crisis (Gunderson 2013). In this chapter, we suggest that environmental sociology and sociological animal studies scholars consider a queering of engagement to a polysemic approach to animalize the subdiscipline to think critically about human, nonhuman, and environmental liberation.

REFERENCES

Abbott, Andrew. 2018. "Varieties of Normative Inquiry: Moral Alternatives to Politicization in Sociology." *American Sociologist* 49 (2): 158–80.

Allen, Michael P., and Erica von Essen. 2018. "Interspecies Political Agency in the Total Liberation Movement." *Between the Species* 23 (1): 7.

Best, Steven. 2014. *The Politics of Total Liberation: Revolution for the 21st Century*. Springer.

Bowers, Melanie M., and Cameron T. Whitley. 2024. "Queer Political Culture in the Face of the Climate Crisis." In *Resolving the Climate Crisis: US Social

Scientists Speak Out, edited by Kristin Haltinner and Dilshani Sarathchandra. Routledge.

Bowers, Robert M., Amy P. Sullivan, Elizabeth K. Costello, Jeff L. Collett, Rob Knight, and Noah Fierer. 2011. "Sources of Bacteria in Outdoor Air across Cities in the Midwestern United States." *Applied and Environmental Microbiology* 77 (18): 6350–56.

Carter, Bob, and Nickie Charles. 2013. "Animals, Agency and Resistance." *Journal for the Theory of Social Behaviour* 43 (3): 322–40.

Carter, Bob, and Nickie Charles. 2018. "The Animal Challenge to Sociology." *European Journal of Social Theory* 21 (1): 79–97.

Cochrane, Alasdair. 2012. *Animal Rights without Liberation: Applied Ethics and Human Obligations*. Columbia University Press.

Dietz, Thomas, Summer Allen, and Aaron M. McCright. 2017. "Integrating Concern for Animals into Personal Values." *Anthrozoös* 30 (1): 109–22.

Dietz, Thomas, Rachael L. Shwom, and Cameron T. Whitley. 2020. "Climate Change and Society." *Annual Review of Sociology* 46 (1): 135–58.

Dunlap, Riley E. 1980. "Paradigmatic Change in Social Science: From Human Exemptions to an Ecological Paradigm." *American Behavioral Scientist* 24 (1): 5–14.

Dunlap, Riley E., and William R. Catton Jr. 1979. "Environmental Sociology." *Annual Review of Sociology* 5 (1): 243–73.

Dunlap, Riley E., and Kent D. Van Liere. 1978. "The 'New Environmental Paradigm.' " *Journal of Environmental Education* 9 (4): 10–19.

Economic Research Service. 2022. "Livestock and Meat Domestic Data." US Department of Agriculture Economic Research Service. https://www.ers.usda.gov/data-products/livestock-and-meat-domestic-data/livestock-and-meat-domestic-data/#Livestock%20and%20poultry%20slaughter.

Flach, Tim. 2017. *Endangered*. Abrams.

Frank, Adam. 2017. "Why I'd Rather Not March." NPR, February 12.

Grauerholz, Liz, Cameron T. Whitley, Erin Kidder, Kelley Ortiz, and Kathy Shepherd Stolley. Forthcoming. "Teaching Human Animal Studies Courses: Exploring Student Characteristics Across Three Universities." *Animals & Society*.

Gunderson, Ryan. 2013. "From Cattle to Capital: Exchange Value, Animal Commodification, and Barbarism." *Critical Sociology* 39 (2): 259–75.

Gunderson, Ryan. 2014a. "The First-Generation Frankfurt School on the Animal Question: Foundations for a Normative Sociological Animal Studies." *Sociological Perspectives* 57 (3): 285–300.

Gunderson, Ryan. 2014b. "Habermas in Environmental Thought: Anthropocentric Kantian or Forefather of Ecological Democracy?" *Sociological Inquiry* 84 (4): 626–53.

Gunderson, Ryan. 2018. "Global Environmental Governance Should Be Participatory: Five Problems of Scale." *International Sociology* 33 (6): 715–37.

Hale, Kayla RS, Fernanda S. Valdovinos, and Neo D. Martinez. 2020. "Mutualism Increases Diversity, Stability, and Function of Multiplex Networks That

Integrate Pollinators into Food Webs." *Nature Communications* 11 (1): 2182.

Halperin, David M. 2003. "The Normalization of Queer Theory." *Journal of Homosexuality* 45 (2–4): 339–43.

Harmon, Amy, and Henry Fountain. 2017. "In Age of Trump, Scientists Show Signs of a Political Pulse." *New York Times*, February 6.

Herrero, Mario, P. Gerber, T. Vellinga, Tara Garnett, A. Leip, C. Opio, H.J. Westhoek, Philip K. Thornton, J. Olesen, and N. Hutchings. 2011. "Livestock and Greenhouse Gas Emissions: The Importance of Getting the Numbers Right." *Animal Feed Science and Technology* 166:779–82.

Irvine, Leslie. 2009. *Filling the Ark: Animal Welfare in Disasters*. Temple University Press.

Jagose, Annamarie. 1996. *Queer Theory: An Introduction*. New York University Press.

Jorgenson, Andrew K. 2018. "Broadening and Deepening the Presence of Environmental Sociology." *Sociological Forum* 33 (4): 1086–91.

Kahn, Brian. 2016. "Scientists Are Planning the Next Big Washington March." *Scientific American*, January 26.

Kalof, Linda, and Amy J. Fitzgerald, eds. 2007. *The Animals Reader: The Essential Classic and Contemporary Writings*. Berg.

Kalof, Linda, and Cameron T. Whitley. 2021. "Animals in Environmental Sociology." In *Handbook of Environmental Sociology*, edited by Beth Schaefer Caniglia, Andrew Jorgensen, Stephanie A. Malin, Lori Peek, David N. Pellow, and Xiarui Huang. Springer.

Koop-Monteiro, Yasmin. 2021. "Including Animals in Sociology." *Current Sociology* 71 (6): 1141–58.

Lacetera, Nicola. 2019. "Impact of Climate Change on Animal Health and Welfare." *Animal Frontiers* 9 (1): 26–31.

MacKendrick, Norah. 2017. "Out of the Labs and into the Streets: Scientists Get Political." *Sociological Forum* 32 (4): 896–902.

Marx, Karl, and Martin Milligan. 1988. *Economic and Philosophic Manuscripts of 1844*. Prometheus Books.

Mattes, Seven, and Cameron T. Whitley. 2021. "Entangled Impacts: Human-Animal Relationships and Energy Development." in *Energy Impacts: A Multidisciplinary Exploration of North American Energy Development*, edited by J.B. Jacquet, J.H. Haggerty, and G.L. Theodori. University Press of Colorado.

McCance, Dawne. 2012. *Critical Animal Studies: An Introduction*. State University of New York Press.

Mori, Kensuke, Melanie Rock, Gavin McCormack, Stefano Liccioli, Dimitri Giunchi, Danielle Marceau, Emmanuel Stefanakis, and Alessandro Massolo. 2023. "Fecal Contamination of Urban Parks by Domestic Dogs and Tragedy of the Commons." *Scientific Reports* 13 (1): 3462.

Morland, Iain, and Dino Willox. 2017. *Queer Theory*. Bloomsbury.

Mortimer-Sandilands, Catriona, and Bruce Erickson. 2010. *Queer Ecologies: Sex, Nature, Politics, Desire*. Indiana University Press.

Nocella, Anthony J., John Sorenson, Kim Socha, and Atsuko Matsuoka. 2014. "Introduction: The Emergence of Critical Animal Studies: The Rise of Intersectional Animal Liberation." *Counterpoints* 448:xix–xxxvi.

Pellow, David Naguib. 2014. *Total Liberation: The Power and Promise of Animal Rights and the Radical Earth Movement*. University of Minnesota Press.

Romo, Amelia B., B. Derrick Taff, Ben Lawhon, Deonne VanderWoude, Peter Newman, Alan Graefe, and Forrest Schwartz. 2019. "Dog Owners' Perceptions and Behaviors Related to the Disposal of Pet Waste in City of Boulder Open Space and Mountain Parks." *Journal of Park and Recreation Administration* 37 (2): 45–64.

Ross, Jeremy. 2017. "Durkheim and the Homo Duplex: Anthropocentrism in Sociology." *Sociological Spectrum* 37 (1): 18–26.

Sbicca, Joshua. 2012. "Eco-Queer Movement (s)." *European Journal of Ecopsychology* 3:33–52.

Seymour, Nicole. 2013. *Strange Natures: Futurity, Empathy, and the Queer Ecological Imagination*. University of Illinois Press.

Singer, Peter. 2012. "Animal Liberation at 30." *Arguing About Bioethics* 185.

Stuart, Diana, and Ryan Gunderson. 2020. "Human-Animal Relations in the Capitalocene: Environmental Impacts and Alternatives." *Environmental Sociology* 6 (1): 68–81.

Sullivan, Nikki. 2003. *A Critical Introduction to Queer Theory*. New York University Press.

Taylor, Nik. 2012. "Animals, Mess, Method: Post-Humanism, Sociology and Animal Studies." In *Crossing boundaries: Investigating human-animal relationships*, edited by L. Birke and J. Hockenhull. Brill.

Taylor, Nik, and Zoei Sutton. 2018. "For an Emancipatory Animal Sociology." *Journal of Sociology* 54 (4): 467–87.

Whitley, Cameron, and Elizabeth Cherry. 2023. "Environmental Sociology and Sociological Animal Studies." In *The Handbook of Teaching and Learning in Sociology*, edited by S. Cabrera and S. Sweet. Edward Elgar Publishing.

Whitley, Cameron T. 2017. "Animals, Humans and Energy Development." In *Fractured Communities*, edited by A.E. Ladd. Rutgers University Press.

Whitley, Cameron T. 2019. "Exploring the Place of Animals and Human–Animal Relationships in Hydraulic Fracturing Discourse." *Social Sciences* 8 (2): 61.

Whitley, Cameron T., and Melanie M. Bowers. 2023. "Queering Climate Change: Exploring the Influence of LGBTQ+ Identity on Climate Change Belief and Risk Perceptions." *Sociological Inquiry*.

Whitley, Cameron T., Erin N. Kidder, Kelley J. Ortiz, and Liz Grauerholz. 2024. "Sociological Animal Studies Courses Are More Effective Than Human-Centered Sociology Courses in Enhancing Empathy." *Teaching Sociology* 52 (4): 309–22.

Whitley, Cameron Thomas, and Linda Kalof. 2014. "Animal Imagery in the Discourse of Climate Change." *International Journal of Sociology* 44 (1): 10–33.

Whitley, Cameron Thomas, Linda Kalof, and Tim Flach. 2021. "Using Animal Portraiture to Activate Emotional Affect." *Environment and Behavior* 53 (8): 837–63.

Wilkie, Rhoda. 2015. "Multispecies Scholarship and Encounters: Changing Assumptions at the Human-Animal Nexus." *Sociology* 49 (2): 323–39.

Woody, Todd. 2021. "Meet the New Climate Refugee in Town: Coyotes." *Bloomberg Green*, December 2.

York, Richard, and Stefano B. Longo. 2017. "Animals in the World: A Materialist Approach to Sociological Animal Studies." *Journal of Sociology* 53 (1): 32–46.

4. Public Sociology for Environmental Health and Justice

Alissa Cordner

When I was in graduate school, I remember coming across Michael Burawoy's (2005) ASA presidential address "For Public Sociology," published just a few years earlier, in a couple of different contexts. In a graduate seminar, we discussed the typology of sociological forms, positioning public sociology generally as a counterpoint to what we were learning to do as PhD students. *Public* sociology was an interesting addition to the *professional* work we were doing, something that could be personally rewarding and useful to others outside sociology but was not the core work of a sociologist.

In a research group, discussions of public sociology were different. I was a member of Phil Brown's Contested Illnesses Research Group (CIRG), which focused on medical conditions with a contested environmental etiology and had partnered with community organizations and nonprofits from its initiation (Brown, Morello-Frosch, and Zavestoski 2012).[1] In this context, discussions of public sociology were about practice and engagement, not theory or typology. For example, when I worked as a research assistant for the community engagement core of Brown University's Superfund Research Core, I didn't wonder whether the discussions I had with community groups or the summaries of technical documents I wrote counted as research or not, or whether the work was more translational or more interactive. It was all part of our work as researchers, and it mattered to us as sociologists because it was useful to the nonacademics with whom we were working.

Today, as a mid-career environmental sociologist, I see deep and meaningful engagement with nonsociologists as something that is both helpful in my research and, at least for my current work, essential if I want my research to matter. If a motivating goal of environmental sociology is to understand how interactions with our environments create and perpetuate inequalities, problems, and crises, much of our work must be directly with the stakeholders and publics who have lived experiences and understandings of those inequalities, problems, and crises.

In this chapter, I argue that environmental sociology is uniquely positioned to contribute not only to scholarly understanding of environmental issues, but to meaningful and justice-oriented action on pressing environmental problems. In particular, a model of *engaged public sociology* allows researchers to collaborate with impacted communities, various other stakeholders, and scholars across many disciplines to develop knowledge that addresses research topics important to those most affected by environmental issues.

WHAT IS PUBLIC SOCIOLOGY?

The roots of **public sociology** are as old as the discipline, far predating its reawakening in the early 2000s (Feagin 2021; London et al. 2024). As Burawoy (2005) himself noted, sociology's earliest theorists "set out to change the world" (5), from Marx's de-alienation of socialism to Durkheim's redemption of organic solidarity (see also Agger 2000). W. E. B. Du Bois was one of the earliest American public sociologists; as Carrera (2023) writes in her analysis of Du Bois's emancipatory sociology, his research had "the intent of transforming the world around him . . ., predat[ing] Burawoy's appeal for public sociology by a hundred years" (10). Similarly, Jane Addams's late nineteenth-century work on urban inequality included the first mixed-method field research in US sociology, and was "used to help local residents understand their community patterns, not just to provide data for publications in academic journals" (Feagin 2021, 25). Thus, engaged and praxis-focused research is not new to the discipline. Rather, it is baked into sociology's inherent attention to inequality and social problems. (This connects to the *implementation environmental sociology* framework that Houser discusses in chapter 2 of this volume.)

Yet for decades, sociology embraced a positivist and instrumentalist approach to research, turning away from topics of policy and public relevance. Starting around 1920, mainstream sociology was focused on "scientific" and "value-free" research, as leading sociologists—mostly white men—pushed the field away from "concerns for social justice and the making of a better society" and toward a "detached and academic sociology" (Feagin 2021, 26, 28; see also Kinloch 1988; Smith 2022; Wilner 1985). While the 1960s saw some resurgence of critical and activist sociology, quantitative and "objective" research remained high-status, with a fear that too much public involvement would "corrupt" scientific research and threaten sociology's external legitimacy and access to resources (Burawoy 2005, 15).

In 2004, when Burawoy delivered his ASA presidential address advocating for public sociology, he was far from the first to talk about outward-facing and engaged sociological research, yet the profile of his speech and the resulting *American Sociological Review* article unquestionably broadened the conversation. Burawoy (2005) dissected the discipline into four categories based on type of intended *audience*—academic or extra-academic—and type of *knowledge* produced—instrumental or reflexive. *Public* sociology—which "brings sociology into conversation with publics" (11)—was distinguished from professional, critical, and policy sociologies. Within public sociology, *traditional* public sociology involves "thin" or "passive" public-facing work by sociologists or journalists that moves sociological research from the academy into the public sphere (Burawoy 2005, 7; see also Gans 2010). This approach has been critiqued as "the sociology of op-ed pages" (Clawson et al. 2007, 5). *Organic* public sociology, on the other hand, relies on a deeper level of engagement, two-way dialogue, and "close connection with a visible, thick, active, local and often counter-public" (Burawoy 2005, 7–8).

Burawoy's very visible definition of public sociology aligned with other conceptualizations of research that was done not just with academics and not just with goals of advancing abstract and supposedly disinterested knowledge. Though definitions differ slightly, public sociology has significant—in some cases complete—overlap with "community-engaged sociology or research, participant action research, community-initiated student-engaged research, service learning, liberation sociology, social justice research, participatory digital sociology," and publicly engaged sociology (Smith 2022, 929). Of course, important distinctions and critiques exist within this general umbrella of approaches. Even Burawoy's impassioned and influential argument in favor of public-facing and engaged work as meaningful and valuable research echoes some of the problems with the mainstream sociology he critiques. First, his identification of the potential "pathology" of public sociology parallels more positivist concerns that too close a connection between the sociologist and publics—particularly *counter* publics—risks derailing research into "faddishness" where the sociological enterprise is "held hostage to outside forces . . . . thereby compromising professional and critical commitments" (Burawoy 2005, 16). Second, his understanding of *publics* emphasizes the public as audience or conversation partner, rather than co-originator or driver of sociological project and thought. As Robert C. Smith noted in his own presidential address—this time to the Eastern

Sociological Society nearly twenty years later (2022)—this framing of "public" ignores other social configurations, particularly that of *community*, and "does not capture the often-close relationships developing via work with community organizations or movements that promote stronger research" (930).

Particularly relevant for my focus on environmental health and justice research, Feagin's (2021) ***critical public sociology*** argues that Burawoy ignores problems with the instrumental positivism that is at the heart of professional sociology. Setting up professional sociology as the sine qua non of all other forms of research (Burawoy 2005, 10) ignores that the social goals (and, I would add, the environmental goals) at the heart of critical public sociology are often in direct conflict with those of mainstream professional sociology (Feagin 2021, 31). Instead of separating critical and public sociology, *liberation sociology* suggests that social justice critiques and goals are central to social science research "concerned about changing and liberating societies away from major types of social oppression" (Feagin 2021, 22; see also Feagin, Vera, and Ducey 2015).

Unlike Burawoy's (2005) model, which is deliberately agnostic about whether public sociology has any normative or political bent (8), social justice is at the heart of this critical public sociology approach. Rather than being concerned that possible politicization might detract from the production of meaningful knowledge about social systems (Smith-Lovin 2007), this approach argues that *all* public sociology should focus on addressing systems of social oppressions with an explicit justice orientation (Feagin, Elias, and Mueller 2009).

Almost twenty years and countless task forces, edited volumes, and funding proposals later, public sociology has gained substantial awareness within, if not influence over, mainstream sociology. ASA President Joya Misra's 2024 presidential address similarly advocated for community-engaged and solutions-oriented sociological practice. Yet the professional costs of doing critical and public sociology are significant (Feagin 2021). From poor reviews on funding proposals, to challenges publishing research findings, to reputational damage, the sociological academy may seem to celebrate public engagement but has yet to fully integrate public sociology into the training of graduate students, the hiring of professors, the institutional evaluation of professional activity, and the reputational prestige of different types of work. Institutionalized systems of promotion, evaluation, and reward fail to recognize publicly engaged work as research, rather than as service (Smith 2022, 926–27). Yet in a political moment characterized by social and political polarization, anti-science rhetoric, and anti-justice attacks from the courts, political institutions, and extremist civil society organizations, public engagement and scholarship to support social justice are more salient than ever (Lamont 2018; Shostak 2018).

WHAT IS ENGAGED PUBLIC SOCIOLOGY, AND WHY SHOULD ENVIRONMENTAL SOCIOLOGISTS CARE?

Moving beyond a one-way dialogue with publics to the type of engaged and justice-oriented research described above, my collaborators and I have argued for a model of ***engaged public sociology*** that works deliberately and deeply with impacted communities throughout the research process (Cordner et al. 2019). Engaged public sociology is an excellent approach for environmental sociologists at any point in their career wanting to do more interdisciplinary, more diverse, and more justice-oriented research. It allows researchers to collaborate with and learn from impacted communities and scholars across many disciplines to develop knowledge that addresses research topics important to those most affected by environmental issues. The public or community can be residents or organizers from a specific geographic community or could be people and communities brought together based on issue rather than location.

This approach is inspired by the principles and practices of **community-based participatory research** (CBPR), a research approach more common in the (environmental) health sciences that involves close, collaborative planning, conduct, application, and translation of research between participants and researchers (Israel et al. 1998; O'Fallon and Dearry 2002; Wallerstein et al. 2017). In CBPR, research participants are involved in the research at every step, from identification of research scope and questions through dissemination of final results and discussions about what to do next. This deep public involvement improves the rigor, relevance, and reliability of research in multiple ways: research questions will be timely and useful; the quality, quantity, and utility of the data collected are improved, particularly in niche or hard-to-access communities; research dissemination and translation are timely and accessible to relevant and necessary audiences; and researchers and communities can each experience increased capacity for future projects (Morello-Frosch et al. 2011). While CBPR is relatively rare in sociology, many social scientists, including sociologists, are moving from a more abstract study *of* environmental and health issues to participation *in* environmental and health research through interdisciplinary collaborations (Matz et al. 2016). Other approaches, including participatory action research (PAR), similarly emphasize the involvement of research participants, particularly those from marginalized populations, and the co-learning and capacity building that results (Fahlberg 2023).

Of course, not all engaged public sociology involves years of on-the-ground collaborations with a grassroots community organization. Indeed, research and research-adjacent activities that meaningfully engage with nonacademic groups and partners can take on many forms, and more rigid definitions or metrics may create unintentional barriers to doing community-engaged research broadly conceived (Key 2019). As an example, while much of my research on toxic chemicals is informed and improved by connections with community groups and activists, my research on wildfire risk management takes different approaches to public connection and involvement. Rather than partner with communities or activists in the design or implementation of my research, I have disseminated my findings—and more general sociological perspectives on wildland fire—to atypical audiences, whether through writing an op-ed about wildfire risk and inequality or leading risk management and communication workshops for prescribed fire trainees (Cordner 2017; Fire Network 2023). I have also prioritized two-way communication with wildland firefighters, fire managers, and land management agency employees, the groups of people I consider to be the most important audience for this research. This two-way communication means that throughout my ethnographic research, I share preliminary findings and seek corrections, expansions, and new examples from participants. Additionally, after papers are published in the peer-reviewed literature, I develop single-page fact sheets to share with practitioners (Cordner 2025).

Engaged public sociology requires a heightened ethical reflexivity, the "self-conscious, interactive and iterative reflection upon researchers' relationships with research participants, relevant communities and principles of professional and scientific conduct" (Cordner et al. 2012, 3). Research done in close engagement or collaboration with communities involves a level of uncertainty not addressed with formal, Belmont-style research ethics covered in an IRB proposal. As colleagues and I described in an article about **reflexive research ethics**, "uncertainty is especially pervasive and salient during research that engages directly with communities and social movements" owing to the emerging nature of technologies and/or methods, significant power relationships between researchers and communities, varying norms across communities, and uneven distribution of research benefits (Cordner et al. 2012, 2). In these types of projects, it is impossible for researchers to anticipate all research decisions and ethical consequences before the research begins. Rather than brainstorming possible ethical issues before research commences and calling it a day, researchers must continually and self-consciously interrogate their practices and relationships to identify and negotiate next steps with an eye to minimizing risks and maximizing benefits. This goes beyond communication of research findings or ideas about "giving back" to a field site (Maiter et al. 2008), and instead demands constant dialogue and interrogation of how researchers' and participants' goals, expectations, and experiences converge or diverge.

Public Sociology and Interdisciplinarity

Doing engaged public sociology often forces—or at least encourages—researchers to move into interdisciplinary areas for the fundamental reason that *neither publics nor environmental problems care about disciplinary boundaries,* however important those boundaries may be to credentialed scholars or however essential scholars think their skills and expertise may be in deciphering, analyzing, and acting on those problems. In particular, a focus on environmental inequality moves researchers away from disciplinary silos and toward trans- and multidisciplinary partnerships with impacted communities and researchers from other fields all working together to answer complicated and multi-scalar issues (see also Houser's chapter in this volume). Social scientists are increasingly involved in collaborations on environmental health topics that require transdisciplinary communication and knowledge (Hoover et al. 2015; Matz, Brown, and Brody 2016). For an environmental sociologist to really engage in scientific and policy debates about air pollution, for example, requires at least some understanding of topics as far-ranging as exposure pathways, emission sources, toxicological mechanisms, epidemiological outcomes, and environmental law and policy.

For example, the PFAS Project Lab, which I codirect with Phil Brown, is a transdisciplinary research group that investigates social and scientific questions related to per- and polyfluoroalkyl substances (PFAS), a class of thousands of widely used toxic chemicals. Our research group includes sociologists alongside scholars from agricultural sciences, anthropology, economics, environmental chemistry, environmental studies, epidemiology, public health, public policy, and toxicology. We engage with and contribute to interdisciplinary environmental health science fields, regularly publishing in journals from sociology and science and technology studies (STS) but also environmental health sciences, occupational health, and public health. This requires us to develop not just interactional but contributory expertise in some of these areas (Collins and Evans 2007). For example, in 2022 we published a paper in an environmental health science methods journal arguing for a new conceptual and operational understanding of likely sources of PFAS contamination (Salvatore et al. 2022), which has already been used by multiple state and federal agencies to better understand potential PFAS contamination.

Similarly, environmental sociologists working on the climate crisis often engage with researchers beyond the social sciences, including climate modelers, ecologists, and physicists. Their work can also contribute directly to the development of climate mitigation and adaptation policy, a clear example of public sociology. For example, at a 2016 federal hearing, multiple US senators used sociological research on climate denial to draw attention to the climate disinformation campaigns and "Web of Denial" funded by fossil fuel companies and their allies (Brulle and Roberts 2017). This public recognition of sociological work was made possible by climate sociologists' intentional engagement with elected officials—for example, through the Climate Social Science Network directed by environmental sociologist Timmons Roberts at Brown University, which supports the work of more than 450 scholars globally doing work on climate politics and obstruction (CSSN 2022).

Public Sociology and Diversity

People of color, Indigenous people, and low-income communities are disproportionately burdened with environmental hazards; this is both something that BIPOC and marginalized communities know and experience firsthand, and something that has been clearly demonstrated in academic research (see Tanesha Thomas's "Intersectionality and Environmental Justice" in this volume). Yet environmental sociology as a field has been intellectually dominated by privileged white scholars, and the field remains a *white space* "characterized by the overwhelming presence of whites in everyday interactions and positions of authority, which formally and informally act as barriers to inclusion and belonging for people of color" (Liévanos et al. 2021, 103). Despite numerous and notable exceptions—particularly from BIPOC environmental sociologists—the field has inadequately engaged with the significance of race and racism, has paid insufficient attention to environmental harms that disproportionately burden people of color, and has failed to center Indigenous values and epistemologies (Liévanos et al. 2021).

Doing public sociology does not necessarily inoculate a researcher from the need to proactively seek to challenge entrenched systems of inequality. As Burawoy (2005) suggested, public sociology "can as well support Christian Fundamentalism as it can Liberation Sociology or Communitarianism"

(8–9). Yet deep and meaningful engagement with publics on environmental issues leads researchers toward recognizing and valuing diverse perspectives and experiences. Thus, it supports efforts to diversify environmental sociology through both areas of research and participation by people with greater embodied diversity and lived experiences of environmental and structural inequalities.

Engaged public sociology also opens up systems of knowledge production to those who lack formal credentials of expertise, focusing on issues and inequalities that matter to impacted and marginalized communities. Knowledge is always produced for someone (Lockie 2017). Engaged public sociology is explicit about producing knowledge for impacted publics, not just for other academics. Thus, this model of research is well-suited to exploring areas of *undone science*, areas where research is not conducted because it does not align with elite priorities (Frickel et al. 2010; Hess 2009), and making visible areas of *unseen science*, research that is conducted but never shared beyond institutional boundaries (Richter, Cordner, and Brown 2018).

Finally, a commitment to public engagement requires researchers to share research and knowledge beyond the academy. In addition to encouraging or even requiring engagement with scholars outside sociology, engaged public sociology can be particularly *accessible* to nonsociologists and nonacademics (Stuart 2021). This may take the form of research translation: writing short, nonacademic summaries of a peer-reviewed paper to share with collaborators and interested organizations, or writing press-releases or op-eds to accompany publications. Other times, making knowledge and research findings accessible leads to the development of new projects or collaborations.

Much environmental justice scholarship could be described as engaged public sociology even when scholars don't claim that label, talking instead about community partnerships or making sure that their research is highly visible beyond the academy. Dorceta Taylor's long-term work on diversity in environmental nonprofits, for example, has had a notable influence beyond the academy through her publications in non-academic venues (e.g., Taylor 2014) and her meaningful engagement with journalists and nonprofit organizations, alongside peer-reviewed papers (e.g., Taylor 2015).

Public Sociology and Justice

Justice is central to public sociology (Carrera 2023; Feagin et al. 2015; Misra 2024; Smith 2022). And just as environmental sociology sees the social and the environmental as inextricable (see Fox, Greiner, and Shtob's introduction in this volume), *social and environmental justice* cannot be separated.

Recognizing that values inevitably influence the research process, environmental sociologists can approach their work with a desire to produce knowledge that supports environmental and social justice, and then do rigorous work toward that goal (Smith 2022, 927).

Because of its motivation to address inequalities through meaningful work with impacted communities, engaged public sociology is inherently justice-oriented. Public sociology is well-suited to investigate larger structures and systems, as Hines (2022) argues in the case of waste as a public sociology issue. This approach is also central to an emancipatory or liberation approach to sociological research, which aims to not only understand but also to *challenge, reduce, and deconstruct* systems of oppression and inequality (Carrera 2023; Feagin et al. 2015). This is particularly relevant when talking about environmental problems, which too often are reduced in media and policy venues to behavioral interventions or technological fixes (Jorgenson 2018). In contrast, a public sociology orientation is attuned to how "structural inequalities and power dynamics from the local to the global contribute to anthropogenic climate change and other environmental problems" (Jorgenson 2018, 1087).

Rather than having an unquestioning acceptance of proposed technocratic solutions, or seeing the abstract and disengaged production knowledge as sociology's number one task, sociologists and other social scientists should work to "document empirically, and ever more thoroughly, the character of major social injustices, nationally and internationally" (Feagin 2021, 36). Of course, as environmental sociologists, we can edit *environmental injustice* into this perspective. Taking a justice-centric, intersectional approach to environmental problems allows researchers not only to better understand multi-scalar and multi-institutional drivers of those problems but to "envision more relevant, multi-scalar, and multifaceted effective solutions to injustice" (Malin and Ryder 2018, 5).

For example, environmental justice scholar Michael Méndez and colleagues have studied climate-induced disasters and social vulnerability, highlighting the impacts of disaster events on incarcerated people, LGBTQIA+ individuals, and undocumented immigrants (Goldsmith et al. 2022; Méndez et al. 2020). His research has involved meaningful partnerships with community groups and impacted residents, pursuing questions of interest to those communities and developing knowledge useful to them, as well as being impactful to policymakers and insightful for other sociologists. He has shared these findings with the press and elected officials, and his research has been influential in making California disaster policy more responsive to the needs of undocumented farmworkers.

THE FUTURE OF ENGAGED PUBLIC ENVIRONMENTAL SOCIOLOGY

If engaged public sociology clearly contributes to a more interdisciplinary, more diverse, and more justice-oriented environmental sociology, must all environmental sociology research take this form? No! This epistemological orientation, like all others, has strengths and weaknesses, and is well-suited to some research questions and methods approaches but not others.

However, I would argue that all environmental sociologists should, in one way or another, engage meaningfully with people, communities, and issues beyond the academy. Whether this means that a nonacademic hobby opens your eyes to a place, people, or problem you hadn't previously considered, or that a conversation with a journalist or policymaker extends your research findings beyond the peer review paywalls, there are many ways for publics to influence your work and for your work to extend to various publics. But for many topics and questions, meaningful public engagement of some kind will make your research better. It also makes it much more likely that the knowledge you develop and the ways you use and share that knowledge can contribute to social and environmental change.

Of course, this type of work carries with it significant barriers, particularly for graduate students, early career professors, or those in more precarious academic appointments. Community-engaged work is often more time- and resource-intensive, and some projects that provide significant benefits to community partners or nonacademic publics offer few of the standard academic benefits in terms of publications, citations, or grants. Reputational gains may be concentrated within members of a specific community, profession, or subculture, rather than among other academic sociologists. Much remains to be done to institutionalize support for engaged community-focused research. Current efforts, including increased grant availability, improved openness at some journals, and support from professional societies, are important first steps. Greater support is needed not just from professional organizations but from graduate advisors and dissertation chairs, tenure and promotion committees, and senior voices in sociology and other social sciences.

Because of environmental sociology's foundational attention to environmental problems and inequalities, our field can benefit significantly from the type of public engagement described in this chapter. It can produce knowledge and contribute to action at multiple scales to understand and address long-standing environmental injustices and to improve public health, advancing more interdisciplinary, more diverse, and more justice-oriented paths forward.

NOTE

1. When Brown moved to Northeastern University in 2012, he started the Social Science Environmental Health Research Institute (SSEHRI), which is still active today. See Social Science Environmental Health Research Institute, Northeastern University, https://www.northeastern.edu/environmentalhealth/. This chapter draws extensively on research I have done as codirector, with Phil Brown, of the PFAS Project Lab, an interdisciplinary research group within SSEHRI focused on environmental health-social science questions about how toxic chemicals are used, regulated, studied, and protested. See The PFAS Project Lab, Northeastern University, https://pfasproject.com/. This chapter is also indebted to a coauthored article on engaged public sociology published in *Environmental Sociology*, the official journal of the International Sociological Association's environmental sociology group (Cordner et al. 2019).

REFERENCES

Agger, Ben. 2000. *Public Sociology: From Social Facts to Literary Acts*. Rowman & Littlefield.

Brown, Phil, Rachel Morello-Frosch, and Stephen Zavestoski. 2012. *Contested Illnesses: Citizens, Science, and Health Social Movements*. University of California Press.

Brulle, Robert J. and J. Timmons Roberts. 2017. "Climate Misinformation Campaigns and Public Sociology." *Contexts* 16 (1): 78–79.

Burawoy, Michael. 2005. "For Public Sociology." *American Sociological Review* 70 (1): 4–28.

Carrera, Jennifer S. 2023. "Advancing Du Bois's Legacy through Emancipatory Environmental Sociology." *Environmental Sociology* 9 (4): 349–65.

Climate Social Science Network. 2022. "About Us." https://cssn.org/about-us/.

Collins, Harry M., and Robert Evans. 2007. *Rethinking Expertise*. University of Chicago Press.

Cordner, Alissa. October 22, 2017. "Opinion: Wildfires like the Wine Country's affect rich and poor differently." *Mercury News*. http://www.mercurynews.com/2017/10/22/opinion-wildfires-like-the-wine-countrys-affect-rich-and-poor-differently/.

Cordner, Alissa. 2025. "Research." https://alissacordner.weebly.com/research.html.

Cordner, Alissa, David Ciplet, Phil Brown, and Rachel Morello-Frosch. 2012. "Reflexive Research Ethics for Environmental Health and Justice: Academics and Movement-Building." *Social Movement Studies* 11 (2): 161–76.

Cordner, Alissa, Lauren Richter, and Phil Brown. 2019. "Environmental Chemicals and Public Sociology: Engaged Scholarship on Highly Fluorinated Compounds." *Environmental Sociology* 5 (4).

Fahlberg, Anjuli. 2022. "Decolonizing Sociology Through Collaboration, Co-Learning and Action: A Case for Participatory Action Research." *Sociological Forum* 38 (1): 95–120.

Feagin, Joe. 2021. "Toward a Critical Public Sociology." In *Routledge International Handbook of Public Sociology*, edited by L. Hossfeld, E.B. Kelly, and C. Hossfeld. Routledge.

Feagin, Joe, Sean Elias, and Jennifer Mueller. 2009. "Social Justice and Critical Public Sociology." In *Handbook of Public Sociology*, edited by V. Jeffries. Rowman & Littlefield.

Feagin, Joe R., Hernan Vera, and Kimberley Ducey. 2015. *Liberation Sociology*. 3rd ed. Paradigm.

Fire Networks. 2017. "Prescribed Fire Training Exchanges (TREX)." https://firenetworks.org/trex/.

Frickel, Scott, Sahra Gibbon, Jeff Howard, Joanna Kempner, Gwen Ottinger, and David Hess. 2010. "Undone Science: Charting Social Movement and Civil Society Challenges to Research Agenda Setting." *Science, Technology, & Human Values* 35 (4): 444–76.

Gans, Herbert J. 2010. "Public Ethnography; Ethnography as Public Sociology." *Qualitative Sociology* 33 (1): 97–104.

Goldsmith, Leo, Raditz, Vanessa, and Méndez, Michael. 2022. "Queer and Present Danger: Understanding the Disparate Impacts of Disasters on LGBTQ+ Communities." *Disasters* 46 (4): 946–73.

Hess, David. 2009. "The Potentials and Limitations of Civil Society Research: Getting Undone Science Done." *Sociological Inquiry* 79 (3): 306–27.

Hines, Myra. 2022. *A Public Sociology of Waste*. Bristol University Press.

Hoover, Elizabeth, Mia Renauld, Michael R. Edelstein, and Phil Brown. 2015. "Social Science Collaboration with Environmental Health." *Environmental Health Perspectives* 123 (11): 1100–1106.

Israel, Barbara A., Amy J. Schulz, Edith A. Parker, and Adam B. Becker. 1998. "Review of Community-Based Research: Assessing Partnership Approaches to Improve Public Health." *Annual Review of Public Health* 19:173–202.

Jorgenson, Andrew K. 2018. "Broadening and Deepening the Presence of Environmental Sociology." *Sociological Forum* 33 (3): 1086–91.

Key, Kent D., Debra Furr-Holden, E. Yvonne Lewis, Rebecca Cunningham, Marc A. Zimmerman, Vicki Johnson-Lawrence, and Suzanne Selig. "The Continuum of Community Engagement in Research: A Roadmap for Understanding and Assessing Progress." *Progress in Community Health Partnerships: Research, Education, and Action* 13 (4): 427–34.

Kinloch, Graham C. 1988. "American Sociology's Changing Interests as Reflected in Two Leading Journals." *American Sociologist* 19 (2): 181–94.

Lamont, Michèle. 2018. "Sociology's Response to the Trump Presidency: Views from the 108th ASA President." *Sociological Forum* 33 (4): 1068–71.

Liévanos, Raoul S., Elisabeth Wilder, Lauren Richter, Jennifer Carrera, and Michael Mascarenhas. 2021. "Challenging the White Spaces of Environmental Sociology." *Environmental Sociology* 7 (2): 103–9.

Lockie, Stewart. 2017. "Post-Truth Politics and the Social Sciences." *Environmental Sociology* 3 (1): s1–5.

London, Rebecca A., Douglas Hartmann, Nancy Plankey-Videla, Elizabeth Borland, and Carol Glasser. 2024. "Community-Engaged Scholarship and Its Implications for Public Sociology and the Discipline." *Social Problems*. December 7. https://academic.oup.com/socpro/advance-article-abstract/doi/10.1093/socpro/spae072/7918801?redirectedFrom=fulltext.

Maiter, S., L. Simich, N. Jacobson, and J. Wise. 2008. "Reciprocity: An Ethic for Community-Based Participatory Action Research." *Action Research* 6 (3): 305–25.

Malin, Stephanie A., and Stacia S. Ryder. 2018. "Developing Deeply Intersectional Environmental Justice Scholarship." *Environmental Sociology* 4 (1): 1–7.

Matz, Jacob, Phil Brown, and Julia G. Brody. 2016. "Social Science-Environmental Health Collaborations: An Exciting New Direction." *New Solutions* 26:349–58.

Méndez, Michael, Genevieve Flores-Haro, and Lucas Zucker. 2020. "The (In)visible Victims of Disaster: Understanding the Vulnerability of Undocumented Latino/a and Indigenous Immigrants." *Geoforum* 116 (1).

Misra, Joya. 2025. " Sociological Solutions: Building Communities of Hope, Justice, and Joy." *American Sociological Review*. https://journals.sagepub.com/doi/10.1177/00031224241302828.

Morello-Frosch, Rachel, Phil Brown, Julia Brody, Rebecca Gasior Altman, Ruthann Rudel, Ami Zota, and Carla Perez. 2011. "Experts, Ethics, and Environmental Justice: Communicating and Contesting Results from Personal Exposure Science." In *Technoscience and Environmental Justice: Expert Cultures in a Grassroots Movement*, edited by G. Ottinger and B. Cohen. MIT Press.

Nature Conservancy. 2017. "Fact Sheet: Prescribed Fire Training Exchanges (TREX)." https://www.conservationgateway.org/ConservationPractices/FireLandscapes/HabitatProtectionandRestoration/Training/TrainingExchanges/Pages/TREX-fact-sheet.aspx.

O'Fallon, Liam, and Allen Dearry. 2002. "Community-Based Participatory Research as a Tool to Advance Environmental Health Sciences." *Environmental Health Perspectives* 110 (S2): 155–59.

Richter, Lauren, Alissa Cordner, and Phil Brown. 2018. "Non-Stick Science: Sixty Years of Research and (in)Action on Fluorinated Compounds." *Social Studies of Science* 48 (5): 691–714.

Salvatore, Derrick, Kira Mok, Kimberly K. Garrett, Grace Poudrier, Phil Brown, Linda S. Birnbaum, Gretta Goldenman, Mark F. Miller, Sharyle Patton, Maddy Poehlein, Julia Varshavsky, and Alissa Cordner. 2022. "Presumptive Contamination: A New Approach to PFAS Contamination Based on Likely Sources." *Environmental Science and Technology Letters* 9 (11): 983–90.

Shostak, Sara. 2018. "Standing up for Science and Social Justice." *Sociological Forum* 33 (1): 242–46.

Smith-Lovin, Lynn. 2007. "Do We Need a Public Sociology?" In *Public Sociology: Fifteen Eminent Sociologists Debate Politics and the Profession in the Twenty-First Century*, edited by D. Clawson, R. Zussman, J. Misra, N. Gerstel, R. Stokes, and D.L. Anderton. University of California Press.

Smith, Robert Courtney. 2022. "Advancing Publicly Engaged Sociology." *Sociological Forum* 37 (4): 926–50.

Stuart, Diana. 2021. *What is Environmental Sociology?* Polity Press.

Taylor, Dorceta E. 2014. "The State of Diversity in Environmental Organizations." Policy Commons. https://policycommons.net/artifacts/1847690/the-state-of-diversity-in-environmental-organizations/2593907/.

Taylor, Dorceta E. 2015. "Gender and Racial Diversity in Environmental Organizations: Uneven Accomplishments and Cause for Concern." *Environmental Justice* 8 (5): 165–80.

Wallerstein, Nina, Nina Duran, John Oetzel, and Meredith Minkler. 2017. *Community-Based Participatory Research for Health: Advancing Social and Health Equity*. Jossey-Bass.

Wilner, Patricia. 1985. "The Main Drift of Sociology between 1936 and 1984." *History of Sociology* 5:1–20.

PART II

Environment, Inequality, and Justice

5. Racial Capitalism and the Environment

Ian Carrillo

What is the relationship between racial capitalism and the environment? Environmental scholars across specializations increasingly interrogate this topic. Gonzalez (2021) uses legal analysis to study the colonial underpinnings of contemporary climate displacement, whereas Táíwò (2021) outlines how historical responsibility for planetary destabilization morally justifies climate reparations. These insights extend decades of racial capitalism research by scholars in the Black Radical Tradition, history, geography, ethnic studies, and the South African anti-apartheid movement (Du Bois 1992; Gilmore 2007; James 1989; Levenson and Paret 2023; Melamed 2015; Robinson 2021; Williams 2021). Environmental sociologists, by possessing a toolkit oriented toward studying social inequalities, human systems, and natural resources, are well-suited to build on this interdisciplinary project. Further, at a time when racial capitalism is of growing importance to general sociology (Bhattacharyya 2018; Go 2021; Itzisgohn and Brown 2020), environmental sociologists can make critical contributions within the discipline.

In examining racial capitalism and the environment, this chapter draws from long-standing research on race and racism. **Race** refers to a group of people who share a common ancestry and physical and cultural traits. Power and hierarchy are intrinsic to the social construction of race, since categorizing someone else is an act of power that often involves placement into a hierarchy (Golash-Boza 2023, 3). For geographer Ruth Wilson Gilmore (2007), **racism** is "the state-sanctioned or extralegal production and exploitation of group-differentiated vulnerability to premature death" (28). Vital to the relationship between racial capitalism and the environment is the process of **racialization**, which involves "the extension of racial meaning to a previously racially unclassified relationship, social practice, or group"

(Omi and Winant 2014, 111). Under racial capitalism, this means bringing ideas about racial differentiation to bear on economic processes and environmental practices.

The term "racial capitalism" originated within the struggle against apartheid in South Africa (Levenson and Paret 2023). Yet it gained a wider reach through the work of Cedric Robinson (2021), whose landmark book *Black Marxism* analyzed how race was critical to the birth and development of capitalism. For Robinson, the formation of racial groups and hierarchies shaped the early foundations of capitalism in Europe, the global expansion of capitalism via European colonialism, and unequal relations in a postcolonial world. Thus, over centuries, white supremacist logic came to structure myriad economic relationships that formed the heart of capitalism, such as work relations, landownership, debt, and trade.

Building on Robinson, work by Laura Pulido put the environment at the forefront of racial capitalism analysis. Pulido (2017) argues that essential to profit-making are racialized environmental inequalities, which materialize from racist ideologies, practices, and state formations. In racial capitalism, environmental racism subsidizes profits and buttresses economic growth, as polluting businesses dump unwanted waste onto racially marginalized and devalued populations (Pulido 2017). These contemporary patterns are a continuation of the primitive accumulation—the dispossession of land, labor, and resources—that defined the process of colonial expansion and produced what Pulido (2018) calls the "racial map of the Anthropocene" (127).

As this chapter details, the environment is central, rather than peripheral, to profit-making and power inequalities in racial capitalism. Extending the productive definitions of racial capitalism that other scholars have provided,[1] and based on my own research in Brazil, I define **racial capitalism** as a process and structure whereby the uneven accumulation of capital occurs through the following: stratification via primarily, but not limited to, racial, class, and gender categories; the theft, depletion, and disorganization of ecosystem resources; economic exchanges centered on labor value expropriation and worker disempowerment; and explicit and implicit support from the state (Carrillo, 2026). This definition differs from past approaches in that it articulates a clearer role for how the environment is configured in the broader relations of racial capitalism.

This chapter describes the utility of the racial capitalism approach for studying problems and debates that have long been important in the environmental sociology canon, ranging from the global political economy to individual actions (Dietz et al. 2020; Klinenberg et al. 2020; Pellow and Brehm 2013; Rudel et al. 2011). I selectively draw from foundational racial capitalism texts and other environmental sociology research. This chapter focuses on the relationship between racial capitalism and the environment at three scales: racial structures and ideologies at the macro level; organizations and institutions at the meso level; and interpersonal and individual dynamics at the micro level. These three scales are useful reference points, as they represent the ideas, practices, and entities that constitute systems of race and racism (Golash-Boza 2016).

I conclude the chapter by discussing how advancing the study of racial capitalism and the environment necessitates not only using approaches within and beyond sociology but also giving attention to the development of theory and praxis oriented around total liberation. For Gilmore (2022), racial capitalism requires building spaces based on curtailed freedom, stolen labor, and depleted ecosystems, all of which structure the unequal accumulation of value. In contrast, she calls for the need for alternative forms of "place-making," whereby people build spaces and communities that prioritize human freedom and autonomy that exist in tandem with vibrant ecosystems.

MACRO-SCALE: RACIAL STRUCTURES AND STRUCTURAL RACISM

For environmental sociologists, the political economy has been an important focal point for understanding environmental change (Rudel et al. 2011). This environmental approach to the **political economy** broadly refers to the struggle between the state, businesses, labor, and civil society over the control of natural resources. A macro-level analysis using racial capitalism shines a light on the racial structures and structural racism in the political economy. It also accounts for how the political economy is nested in a world system stratified through colonial projects that use ideas of white supremacy and racial difference to devalue nonwhite people and territories. Walter Rodney (1981) undertakes such an analysis in *A History of the Guyanese Working People, 1881–1905*. This work provides three interconnected approaches that are useful for environmental sociologists: the linkages between environmental transformation and capitalist development; the entanglements between environmental processes and racialized class relations; and the influence of a colonial world system on local racialized class interests.

First, the relationship between environmental transformation and capitalist development is central to racial capitalism. In Guyana, sugarcane production—the main economic activity—operated under major environmental constraints, as plantations were located on a thin coastal strip prone to flooding and drought. Producers transformed the environment by building elaborate systems of dams, dykes, and canals to suit production demands. Guarding against and recovering from flooding was a constant issue, since producers created areas for cultivation by draining water from land that sat at or below sea level. Many plantations had two dams: one to keep out salt water and another to let in fresh water. Producers built canals as tools for irrigation, drainage, and transport. Sluices were installed in dams to regulate water levels in the canals. Overall, this duality of environmental constraint and transformation, centered around the "sea-defense problem," defined many of the dilemmas, debates, and decisions in Guyana's political economy (Rodney 1981, 2).

The second analytic area highlights how racialized class relations are intertwined with environmental processes. In Guyana's plantation society, as Rodney (1981) puts it, "generations of blacks working under white masters had markedly transformed this coastal habitat" (1). White planters used a mix of enslaved, indentured, and free laborers recruited from Africa and India. These non-European laborers toiled in environmentally hazardous conditions, as they moved tons of soil, clay, mud, and water to create usable farmland and canals. Producing sugarcane was also physically exhausting and dangerous work, involving cane cutting, loading, transport, distilling, and planting.

The white planter class wielded power on plantations and in the political economy. For the former, planters determined work routines and conditions, which included task assignments, remuneration, workday length, and health and safety. In the political economy, planter elites populated most legislative and administrative offices, where defending racialized class interests guided their decision-making. Although the sea-defense problem afflicted all of society, planters' management of the issue was frequently self-serving. For example, white elites directed the government funds used to finance dams, drainage pumps, and sluices to their own plantations, thereby leaving workers' villages and subsistence farming areas to languish under wet conditions. Even in moments of extreme inundation and drought, planters hoarded key resources, like drainage pumps and fresh water, deliberately increasing rates of disease and death among the nonwhite population. Overall, this racialized class structure not only anchored the relationship between labor, capital, and the state, but also the management of environmental resources.

The third approach focuses on how local racialized class interests are nested in a global political economy borne from colonialism. Under the yoke of Great Britain, Guyana's status as a colony limited the options and actions of planter elites in numerous ways. With most capital in British hands, Guyanese elites had little financial power to invest in science and technology to improve production strategies. Moreover, the high costs of technology related to flood management, such as drainage pumps, resulted in Guyana holding extensive capital debt to British financiers. Guyana was thus stuck in a financial dilemma, common among colonized nations, that was difficult to escape.

This colonized status created deep vulnerabilities for Guyana, as it made the country's interests subordinate to the domestic interests of colonizer nations. For example, Germany's heavily subsidized beet sugar was sold to British consumers at a cheaper price than Guyana's cane sugar. British leaders opted to not confront Germany about its dumping practices, since they saw political advantage both in low food prices for their domestic population and in cheap raw material inputs for the nation's emergent candy industry. As financial firms in London consolidated market share and political power, they began allocating capital based on profit rates and return on investment, which further restricted Guyana's access to capital. Guyana's colonized status even engendered difficulties as they sought to create new economic and trade partnerships. For instance, seeing economic opportunities for sugar exports dwindling in Great Britain, Guyana began exporting sugar to the United States. Yet Guyana needed Britain to grant permission to do so, a process that required extensive negotiations, while their subordinate status to the United States imbued dependency in this new relationship.

These global trade dynamics had powerful impacts locally. Guyana's working population faced great hardships from international pressures. As British financial firms prioritized returns on investment, working people in Guyana increasingly confronted exploitation and neglect. Workers experienced worsening labor conditions and lower wages. New agricultural machinery displaced workers, increasing unemployment and reducing job opportunities throughout the economy. Overall, Rodney (1981) saw domestic and international constraints as deeply entangled: "Operationally, the two were often merged, because dependency was not simply a matter of foreign trade but was reflected in and supported by the internal socioeconomic structures" (29–30).

These analytic tools are useful for environmental sociologists studying racial capitalism at the macro-scale. Structural racism and racial structures define the mutually constitutive relationship between the political economy and environmental transformation. Recent research showcases other contexts for a similar macro-scale analysis. Studying colonialism in the US

Northeast, Murphy (2021) highlights the process by which European settlers displaced Native populations and enacted new agricultural practices, thereby altering ecosystems, increasing emissions, and raising temperatures. McGee and Greiner (2020) detail how the development of the fossil fuel economy in the United States depended on the sustained exploitation of Black workers via slavery, sharecropping, and two-tiered industrial employment. The racial capitalism approach thus has the potential to make novel contributions to areas of study that are of long-standing and contemporary interest to environmental sociologists.

MESO-SCALE: ORGANIZATIONS AND INSTITUTIONS

In environmental sociology, meso-scale analysis often focuses on corporate behavior, corporate culture, and environmental politics within and across sectors (Dietz et al. 2020). This work has mostly been race-neutral, yet it is increasingly clear that race and racialization play powerful roles in meso-level behavior (Ray 2019). This section of the chapter, divided into two parts, outlines pathways for racial capitalism to advance these areas of inquiry. I first use David Roediger and Elizabeth Esch's (2012) *The Production of Difference* to discuss the activities of organizations directly responsible for environmental despoiling. The second part draws from Jill Harrison's (2019) *From the Inside Out* to examine the state institutions defending organizations that destroy the environment. This analysis reveals how meso-scale dynamics, such as organizational logic, fields, and rationality, are vital for sustaining the relationship between racial capitalism and the environment.

Roediger and Esch (2012) describe the frontline processes whereby race, class, and the environment intersected in capitalist development and colonial settlement in the United States. By focusing on worksite activities, they illustrate how racial ideologies were imbued in bureaucratic rationality and organizational behavior in key sectors responsible for environmental transformation, such as steel, auto, coal, meatpacking, railroads, and dam and canal construction. They stress how processes of racialization were central to the labor and natural resource management that buttressed capital accumulation. On these worksites, racialization permeated the feedback loop between cultural ideas, such as expectations, roles, identities, and labor conditions, comprised of wages, tasks, rights, and health and safety.

Bureaucratic rationality operated in three key ways. The first entailed using managerial science to promote and justify racial difference. Roediger and Esch (2012) detail how leading industry journals propagated racist ideas to guide labor management strategies. For instance, industry psychology spread the eugenicist notion that race and IQ were related, while other managerial science publications embraced the myth that racial groups had innate differences related to work ethic and physical traits. These ideologies crucially provided scripts and templates that industry leaders and managers used to link racial identity with labor roles and expectations.

The second area involved the enactment of so-called racial difference in worksites and environmentally hazardous labor processes. Race-making went in hand in hand with environmental change, as white managers encoded workers as racially superior and inferior, and ensured that those hierarchies were reflected in the relative environmental danger of work assignments. Such patterns existed across multiple sectors. In steel and auto, managers steered Mexican, Black, and European ethnic immigrants into the hottest and most unsafe jobs, such as work with blast furnaces, chemicals, and foundries. Mexican and Black workers labored on the killing floor and with fertilizer in meatpacking, while also occupying the wettest jobs in the stockyards. In mining, managers gave the most dangerous above-ground jobs, such as smelting, to Mexicans. In the Panama Canal construction, Afro-Jamaicans, Irish, and Chinese workers were assigned digging tasks, where death and suicide rates were very high (see fig. 5.1). In all these positions, supposed racial inferiority and environmental danger also corresponded with low wages and weak protections.

At the same time, the hierarchical position of white workers was defined in opposition to the expendability and devaluation of nonwhite workers. Managerial science argued that northern Europeans were not naturally suited for work in high temperatures, and that white workers were predisposed to excel at higher-skilled positions, like kiln-building, as opposed to placing bricks in kilns, which was reserved for Black workers. Importantly, administrative science stated that only whites were innately suited for occupying managerial positions and for commanding workers seen as inferior. Furthermore, since whites were seen as "the best enforcers of Jim Crow," the cultural expectation was for whites to wield power at the top of the hierarchy, with nonwhite workers expected to follow orders (Reedier and Esch 2012, 192).

Lastly, white managers deliberately used ideas of racial difference to undermine worker solidarity. One major strategy was for managers to create racially sorted work gangs, thereby exploiting the "natural rivalries" between workers that managerial science argued was a racial reality (Roediger and Esch 2012, 154). This racial stratification aimed to deter workers from recognizing any cross-racial class interests. Managers used

FIGURE 5.1. Workers toil with pickaxes on a ridge to clear space for the Panama Canal. Source: Underwood & Underwood, 1904. Caption below photograph reads: "Workmen digging away a mountain, where the great Canal cuts through the Culebra ridge, Panama." Retrieved from the Library of Congress, https://www.loc.gov/item/2021639845/.

other racial strategies to sow resentment and distrust between workers. For instance, as environmental dangers permeated these worksites, managers racialized the issue of safety. When environmental accidents—for example, a mining collapse or explosion—occurred, white managers strategically blamed such catastrophes on non-white workers, saying that their supposedly inferior racial characteristics were responsible for failing to uphold safety protocols. In other cases, if white workers began organizing, managers often placed Black workers in more skilled positions to show white workers they were replaceable. If white workers went on strike, industry leaders exploited Black and Mexican workers' socio-economic vulnerability by using them as strike-breakers, thereby deepening whites' distrust and animosity toward workers of color. Overall, this racial stratification pitted workers against each other in the service of suppressing wages and keeping worksites unsafe, while also ensuring that racialized capital accumulation and environmental transformation proceeded uninterrupted.

The state institutions that defend environmentally exploitative businesses are another crucial meso-level aspect of racial capitalism. Harrison's (2019) institutional ethnography of the US Environmental Protection Agency (EPA) provides a rich study of the relationship between racialization, environmental harm, and bureaucratic rationality. This organizational analysis of the EPA's failure to uphold laws against environmental racism offers two insights on the "state-sanctioned violence" that sustains racial capitalism (Pulido 2017).

First, the bounded rationality internal to the EPA and inculcated in bureaucrats contributed to the EPA's neglect of minoritized communities in multiple ways. Seeing that polluting companies wielded disproportionate political-economic power, bureaucrats were fearful of raising the ire of public and private sector elites, who could file expensive lawsuits, cut budgets, and imperil operating resources in general. By acquiescing to polluter interests, bureaucrats often failed to enforce environmental justice laws and frequently approved environmental permit requests. Bureaucrats recognized that they could endanger their career by challenging the notion that industry interests superseded community well-being. This dynamic was further exacerbated by the presence of former industry actors in EPA leadership positions. In this context, even the bureaucrats committed to environmental justice initiatives saw their projects sidelined and their funding diminished. While this bounded rationality was seemingly nonracial, it nevertheless guided organizational behavior that defended the industrial activities responsible for environmental racism.

Second, colorblind narratives were imbued in organizational logic and identity to downplay racial disparities and obscure racist legacies of environmental inequalities. Many EPA bureaucrats embraced the idea that bureaucratic race-neutrality was a guiding principle, central to the EPA's organizational identity, and core to their identity as EPA bureaucrats. As a consequence, bureaucrats commonly felt that environmental justice projects were contrary to the EPA's role and purpose. Many bureaucrats preferred using quantitative metrics considered unbiased and more "scientific," and rejected qualitative, historical, and community-led methods that better captured the real-world dynamics of environmental racism. The experiences and views of white male bureaucrats who denied the gravity of environmental racism were overvalued and excessively influential. Bureaucrats frequently disparaged environmental justice staff, labeled such projects pejoratively as "reverse racism," and blamed minoritized communities for their plight. These colorblind approaches served to not only obscure the racist realities of industrial pollution but also to delegitimize the aims of the bureaucrats and communities fighting against racial inequalities.

In sum, there are several meso-scale approaches for analyzing racial capitalism and the environment. Focusing on organizations driving environmental change, Roediger and Esch (2012) elucidate how racial ideas converge with business strategies to activate the hierarchies and environmental transformation undergirding racialized capital accumulation. Harrison's (2019) examination of the state institutions defending industrial pollution reveals that such entities often engage in systematic neglect, denialism, and obstruction of racial justice. Overall, this scholarship unearths potential pathways for studying institutional behavior, organizational culture, and environmental politics through a racial capitalism lens.

MICRO-SCALE: INTERPERSONAL AND INDIVIDUAL DYNAMICS

How emotions, values, beliefs, and attitudes guide individual decision-making related to environmental behavior has long been a focus in environmental sociology (Dietz et al. 2020). Race scholars similarly stress that interpersonal and individual dynamics are important for creating and sustaining institutional and systemic racism (Golash-Boza 2016). To examine how the relationship between racial capitalism and the environment operate at the micro-scale, this section draws from two studies. One is Nadia Kim's book, *Refusing Death* (2021), which investigates how Latina and Asian women in Los Angeles built an environmental justice movement. The other is my own research on white male elites in environmentally hazardous firms in Brazil (Carrillo 2021). These studies show how race, class, and gender, among other social categories, intersect to shape emotions and influence how power is both challenged and wielded.

Kim (2021) analyzes how industrial pollution, emanating from ports, railyards, diesel trucks, and power plants, negatively impacts the lives of working-class and immigrant communities in Los Angeles. In studying the women who mobilize to push back against many forms of environmental injustice, she interrogates how emotions operate as a key mode of power, authority, and counter-hegemonic subversion. The micro, therefore, is entangled with the meso and the macro: "Emotions . . . are always social structures that connect individuality to the collective and are structures of power embedded in social systems, institutions, and cultures, not mere ephemeral forms of visceral embodiment" (Kim 2021, 99).

This analysis builds on research on the racialized, classed, and gendered dynamics of emotions. For Eduardo Bonilla-Silva (2019), racialized emotions are vital elements of intergroup relations, individuals' subjectivity, and perceptions of whose feelings are seen as legitimate or illegitimate. Taking a feminist, antiracist approach, Kiran Mirchandani (2003) states that race, class, and gender codetermine roles and expectations for emotion work. This intersectional frame underscores how low-income women of color bear distinct burdens for paid and unpaid emotion labor, and that racial formation is also a "gendered, sexualized, and classed" process (Mirchandani 2003; Kandaswamy 2012, 27). As Gargi Bhattacharyya (2018) argues, the domestic work and care work that women of color do across the globe undergirds the social reproduction of racial capitalism and reflects the intrinsic function of patriarchy in racial capitalism.

Emotions play three major roles in the context of Kim's study. Powerful actors wielded emotions to defend environmentally harmful practices necessary for business growth. In Los Angeles, government and corporate officials used emotions to delegitimize community concerns in public meetings to discuss local environmental issues. In these meetings, officials appeared visibly irritated, openly rolled their eyes, and performatively sighed when activist women communicated their concerns about pollution. Officials utilized economic threats to incite fear, which had the effect of potentially silencing low-income community members. In sum, these officials used their emotions to dismiss critique, defend the status quo, and protect power against challenges from below.

Second, immigrant women employed a range of emotive strategies to combat officials' emotional indifference and apathy. Women highlighted their role as maternal figures seeking to protect children from environmental danger, thus appealing to society's cultural sympathy with maternal love. Deliberately seeking to gain attention by violating feminine expectations, they gave angry speeches. In some cases, they appealed to traditional femininity by smiling and speaking in polite, soft tones. Women attempted to shame and guilt officials into becoming sympathetic with their plight. In interviews, these women described engaging in processes of active learning and experimentation, as they consciously evaluated which emotional strategies worked best to advance their cause. Overall, in this context of extreme power imbalances, emoting was one of the few tools available for immigrant women to use in order to challenge hegemonic authority.

The third role is emotional labor. For immigrant women, this involved not only coping with the general stress of living in proximity to pollution but also doing the care work for people suffering from cancer, asthma, and respiratory disease after prolonged toxic exposure. Environmental justice activism and care work became bound together in the same project, as women acted collectively to improve environmental amenities in their community, while also attending to the health needs of sick children and elderly people at home. This unpaid labor was in constant demand, thereby representing another mechanism through which racial capitalism extracts and expropriates value to subsidize profit accumulation.

In contrast to analyzing communities where businesses externalize unwanted costs and materials, I look inside the hazardous companies themselves to understand how emotions and attitudes shape environmental decision-making (Carrillo 2021). I study the white male elites who own and manage Brazilian sugar-ethanol mills, where racialized class stratification is central to labor and environmental activities. My interviews with these men reveal how racialized emotions and group-based relations (Bonilla-Silva 2019) guided their environmental decision-making. Two examples show how white male elites wield emotive power to legitimize the subjugation of Black and Brown workers in environmentally hazardous labor, such as cane cutting and straw burning (see fig. 5.2).

First, elites express beliefs and hold attitudes reflecting their view that the world is comprised of superior and inferior groups. Elites categorize themselves in the former, and thus see themselves as beacons of morality, yearn for their power to be seen as natural and noble, and expect to be obeyed. Inversely, elites view workers of color as inherently lazy, unintelligent, and unskilled. Such beliefs follow from colonial systems of racialization, whereby sugarcane work and workers were encoded as Black (Silva 2016), with racial stereotypes used to justify class exploitation. An industry consultant described his superior attitude toward workers:

> In Brazil, tales of suffering are appealing. Success is routinely shamed, and profit is seen as a shameful pursuit, from the perspective of low-skilled workers . . . Rural workers have little comprehension of the future. Rural workers can't even comprehend months ahead of time. For them, daily or weekly time horizons are more manageable . . . The worker is scared of being the boss because his team is his brother, his

FIGURE 5.2. Cane-cutters work in a field of sugarcane at a mill in Northeast Brazil. Photo credit: Ian Carrillo.

> neighbor, his cousin . . . Workers can't align behavior with cognition. They routinely do things that they know are wrong. For instance, they are like a diabetic eating sugary products, knowing he shouldn't. Their behavior overrides comprehension. He just does what he likes. (Carrillo 2021, 62)

This emotive power guides everyday business activities and profit-making, as white elites deliberately steer workers of color into brutal labor and environmental tasks. Furthermore, this attitude of perceived superiority legitimizes the expropriation of value from workers and nature.

Second, elites stoke fear and anxiety to undermine attempts at improving environmental practices. This was evident in the way white elites responded to efforts by public agencies and grassroots activists to prohibit straw burning. Mills incinerate straw so that workers have easier access to the cane stalks with their machetes. While straw burning substantially increases worker productivity, it also pollutes the air of nearby communities, emits carbon into the atmosphere, and imperils worker safety. The emergence of laws and public demands to eliminate straw burning alarmed elites. Seeking to subvert these pressures, elites spread the idea that environmental restrictions would reduce mill profitability and bankrupt business operations. One mill director made this point by stoking racialized panic:

> And the crime, the hunger, the robberies, the misery, it will all unfold, because the security belt for [the state capital] Recife is the sugar zone. If the mills close, there isn't enough police in the world for that. Everyone would be killed, hundreds killed, dozens every day. (Carrillo 2021, 61)

This respondent raises the specter of the racialized violence and mass chaos that would break out if mills were to close. In fanning such racial anxiety, he aims to delegitimize an environmental justice movement that might threaten mills' profit margins. In doing so, he perpetuates the historical stereotype that Black and Brown workers are innately criminal, that mills are a key social instrument keeping workers under control, and that a society without mills' disciplinary pressure would be gravely unsafe.

Overall, these studies elucidate micro-scale features in the relationship between racial capitalism and the environment. To investigate how power operates from below, Kim (2021) studies the low-income Latina and Asian immigrant women who lead a grassroots environmental justice movement. Inversely, I interrogate how white male elites, as owners and managers of hazardous companies, wield power from above (Carrillo 2021). Such approaches map onto environmental sociology research focusing on emotions, values, and attitudes. Yet they also connect such micro-level dynamics to the institutions, structures, and systems through which racial capitalism operates to create environmental inequalities and injustices.

CONCLUSION

This chapter outlines potential pathways for examining the linkages between racial capitalism and the environment. Launching such a research agenda requires navigating interdisciplinary and disciplinary approaches. On the one hand, racial capitalism largely originated and developed outside the discipline of sociology (as an exception, see Du Bois 1992, 2007). Environmental sociologists must look beyond our canon to study socio-environmental systems through a racial capitalism framework. On the other hand, a rich set of debates in environmental sociology focus on the problems, dilemmas, and potential solutions that permeate socio-environmental systems. Rather than abandon these debates, environmental sociologists can bring a racial capitalism analysis to bear on these bedrock areas of the sub-field. This chapter provides a (limited) template for how environmental sociologists might implement such an approach in future work.

Importantly, leading scholars of racial capitalism stress that the framework involves more than studying the myriad entanglements between oppression and capitalism: it should also generate possibilities for theorizing and practicing total freedom (Kelley 2023). Ruth Wilson Gilmore (2022) calls this project "abolition geography":

> [It] starts from the homely premise that freedom is a place. Place-making is normal human activity: we figure out how to combine people, and land, and other resources with our social capacity to organize ourselves in a variety of ways, whether to stay put or to go wandering. Each of these factors—people, land, other resources, social capacity—comes in a number of types, all of which determine but do not define what can or should be done. Working outward and downward from this basic premise, abolitionist critique concerns itself with the greatest and least detail of these arrangements of people and resources and land over time. (74–75)

We see examples of contemporary communities working toward these alternative forms of "place-making." In Brazil, *quilombos*—communities historically settled by Afro-Brazilian and Indigenous people that escaped from slavery—combine land autonomy and ancestral knowledge to build sustainable social and agricultural systems (Silva 2023). Anchored in solidarity and cooperation, these values and practices resemble the concepts of *buen vivir* in Andean Indigenous communities in South America and of *ubuntu* in Xhosa and Zulu communities in South Africa (Gudynas 2011; Ramose 2010; Silva 2023). Thus, when seeking strategies to resist racial capitalism, there is already a rich legacy of human traditions from which to draw.

To conclude, racial capitalism research can pursue twin objectives. One is to understand how repressed groups dislodge hegemonic systems and enshrine emancipatory epistemologies and practices into new structures and systems (Táíwò 2021). The other is to decolonize the discipline of sociology, with the aim of shedding the legacies of coloniality in the methods, theories, and debates that populate the field (Meghji 2020). Similarly, in their chapters in this book, Jules Bacon and Kirsten Vinyeta (chapter 6) and Michael Warren Murphy (chapter 7) outline how critically interrogating histories and realities of empire and colonialism are imperative for decolonial approaches to environmental sociology. For environmental sociologists, these tasks are not just long overdue; they are also necessary to meet the urgent challenges arising from intensifying environmental and climate crises across the globe.

NOTE

1. See Bhattacharyya 2018; Foster and Clark 2020; Gilmore 2007; Jenkins and Leroy 2021; and Pulido 2017.

REFERENCES

Bhattacharyya, Gargi. 2018. *Rethinking Racial Capitalism*. Rowman & Littlefield.

Bonilla-Silva, Eduardo. 2019. "Feeling Race: Theorizing the Racial Economy of Emotions." *American Sociological Review* 84 (1): 1–25.

Carrillo, Ian. 2021. "Racialized Organizations and Color-Blind Racial Ideology in Brazil." *Sociology of Race and Ethnicity* 7 (1): 56–70.

Carrillo, Ian. 2026. *The Business of Racism: Labor and Environment in Brazil's Racial Capitalism*. Duke University Press.

Dietz, Thomas, Rachael Shwom, and Cameron Whitley. 2020. "Climate Change and Society." *Annual Review of Sociology* 46 (1): 135–58.

Du Bois, W.E.B. 1992. *Black Reconstruction in America*. Free Press.

Du Bois, W.E.B. 2007. *The World and Africa—Color and Democracy*. Oxford University Press.

Foster, John Bellamy, and Brett Clark. 2020. *The Robbery of Nature*. Monthly Review Press.

Gilmore, Ruth Wilson. 2007. *Golden Gulag*. University of California Press.

Gilmore, Ruth Wilson. 2022. *Abolition Geography*. Verso.

Go, Julian. 2021. "Three Tensions in the Theory of Racial Capitalism." *Sociological Theory* 39 (1): 38–47.

Golash-Boza, Tanya. 2016. "A Critical and Comprehensive Sociological Theory of Race and Racism." *Sociology of Race and Ethnicity* 2 (2): 129–41.

Golash-Boza, Tanya. 2023. *Race and Racisms, Brief 3rd Edition*. Oxford University Press.

Gonzalez, Carmen. 2021. "Racial Capitalism, Climate Justice, and Climate Displacement." *Oñati Socio-Legal Series* 11 (1): 108–47.

Gudynas, Eduardo. 2011. "Buen Vivir: Germinando Alternativas al Desarrollo." *América Latina en Movimiento, ALAI* 462:1–20.

Harrison, Jill Lindsay. 2019. *From the Inside Out*. MIT Press.

Itzisgohn, José, and Karida Brown. 2020. *The Sociology of W.E.B. Du Bois*. NYU Press.

James, C.L.R. 1989. *The Black Jacobins*. Vintage Books.

Jenkins, Destin, and Justin Leroy. 2021. "Introduction: the Old History of Racial Capitalism." In *Histories of Racial Capitalism*, edited by Justin Leroy and Destin Jenkins. Columbia University Press.

Kandaswamy, Priya. 2012. "Gendering Racial Formation." In *Racial Formation in the Twenty-First Century*, edited by Daniel Martinez HoSang, Oneka LaBennett, and Laura Pulido. University of California Press.

Kelley, Robin. 2015. *Hammer and Hoe*. University of North Carolina Press.

Kelley, Robin. 2023. "Why We Are Called Hammer & Hope." *Hammer & Hope* 1:4.
Kim, Nadia. 2021. *Refusing Death*. Stanford University Press.
Klinenberg, Eric, Malcolm Araos, and Liz Koslov. 2020. "Sociology and the Climate Crisis." *Annual Review of Sociology* 46:649–69.
Levenson Zachary and Marcel Paret. 2023. "The South African Tradition of Racial Capitalism." *Ethnic and Racial Studies* 46 (16): 3403–24.
McGee, Julius, and Patrick Greiner. 2020. "Racial Justice Is Climate Justice: Racial Capitalism and the Fossil Fuel Economy." Hampton Institute. PDXScholar. https://pdxscholar.library.pdx.edu/usp_fac/278/.
Meghji, Ali. 2020. *Decolonizing Sociology: An Introduction*. Polity.
Melamed, Jodi. 2015. "Racial Capitalism." *Critical Ethnic Studies* 1 (1): 76–85.
Mirchandani, Kiran. 2003. "Challenging Racial Silences in Studies of Emotion Work: Contributions from Anti-Racist Feminist Theory." *Organization Studies* 24 (5): 721–42.
Murphy, Michael. 2021. "On the Eco-Materiality of Racial-Colonial Domination in Rhode Island." *Political Power and Social Theory* 38: 161–89.
Omi, Michael, and Howard Winant. 2014. *Racial Formation in the United States*. Routledge.
Pellow, David, and Hollie Nyseth Brehm. 2013. "An Environmental Sociology for the Twenty-First Century." *Annual Review of Sociology* 39:229–50.
Pulido, Laura. 2017. "Geographies of Race and Ethnicity II: Environmental Racism, Racial Capitalism and State-Sanctioned Violence." *Progress in Human Geography* 41 (4): 524–33.
Pulido, Laura. 2018. "Racism and the Anthropocene." In *Future Remains*, edited by Gregg Mitman, Marco Armiero, and Robert Emmett. University of Chicago Press.
Ramose, Mogobe. 2015. "Ecology Through Ubuntu." In *Environmental Values*, edited by Roman Meinhold. Konrad-Adenauer Stiftung.
Ray, Victor. 2019. "A Theory of Racialized Organizations." *American Sociological Review* 84 (1): 26–53.
Robinson, Cedric. 2021. *Black Marxism: The Making of the Black Radical Tradition*. University of North Carolina Press.
Rodney, Walter. 1981. *A History of the Guyanese Working People, 1881–1905*. Johns Hopkins University Press.
Roediger, David, and Elizabeth Esch. 2014. *The Production of Difference: Race and The Management Of Labor In U.S. History*. Oxford University Press.
Rudel, Thomas, J. Timmons Roberts, and JoAnn Carmin. 2011. "Political Economy of the Environment." *Annual Review of Sociology* 37 (1): 221–38.
Silva, Maria. 2016. "Trabalho Rural: As Marcas da Raça." *Lua Nova* 99: 139–67.

Silva, Silvane. 2023. "Fighting to Preserve Black Life and Land Rights: A Study of Quilombolas in the State of São Paulo, Brazil." In *Beyond Racial Capitalism*, edited by Caroline Hossein, Sharon Austin, and Kevin Edmonds. Oxford University Press.

Táíwò, Olúfẹ́mi O. 2021. *Reconsidering Reparations*. Oxford University Press.

Williams, Eric. 2021. *Capitalism and Slavery*. University of North Carolina Press.

6. Contesting the Settler Structure

Indigeneity, Settler Colonialism, and Environmental Sociology

J.M. Bacon and Kirsten Vinyeta

As sociologists, we sometimes describe our scholarly role as making the invisible *visible*, since we unveil the inner workings of social structures that, over time, become so normalized and taken for granted they reach a point of naturalization and invisibility. **Settler colonialism** is one such structure—so normalized and invisible to the settler populace, in fact, that even sociologists whose job it has been to reveal the forces covertly structuring society had failed to meaningfully engage the concept until the last fifteen years. Like *Fight Club*, Lorenzo Veracini (2011) might argue that for settlers *the first (and second!) rule of settler colonialism is you do NOT talk about settler colonialism.* Yet, in countries like the United States, Canada, and Australia, settler colonialism is a pervasive social structure that—in shaping social relations and hierarchies between settler states, Indigenous and non-Indigenous peoples—drives dominant environmental practices and generates myriad, intertwined environmental injustices (see fig. 6.1).

Because, as many have theorized, settler colonialism is an eco-social structure principally concerned with land acquisition and control, it inherently undergirds all environmental questions in settler nation-states. This is not to suggest that settler colonialism is the only salient structure for understanding environmental issues, but it is a structure that must be considered within the complex web of social relations that Patricia Hill Collins (2022) refers to as the matrix of domination (for more on intersectionality see Tanesha A. Thomas's chapter in this volume). Settler colonialism exerts intersectional, oppressive logics and consequences. In its rejection of Indigenous identities, communal structures, land tenure, and ecologies, and in its institutionalization of dominant European institutions such as capitalism, racism, heteropatriarchy, Christianity, and later, Western science, settler

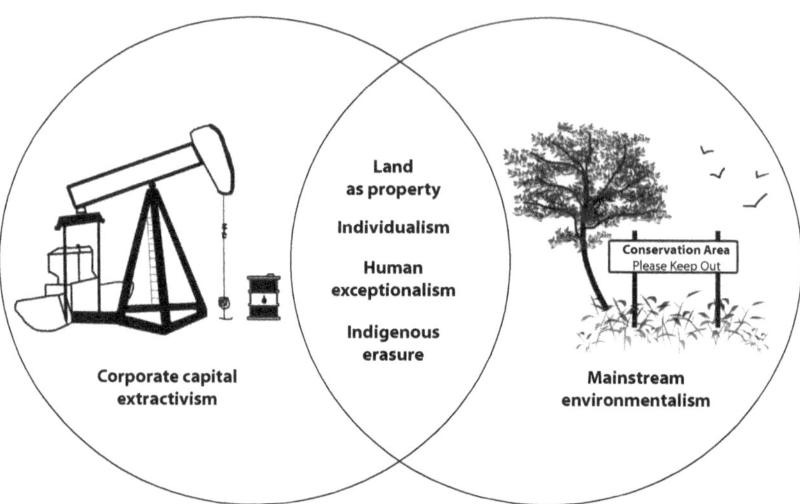

FIGURE 6.1. Shared eco-social norms under settler colonialism. Source: authors.

colonialism weaves land dispossession together with the disempowerment of women, the large-scale exploitation of more-than-human species, and the outcasting of ways of knowing and being that disrupt settler dominance. As such, settler colonial and Indigenous studies are powerful lenses through which environmental sociologists can (1) better understand and dismantle intersectional environmental injustices unfolding in settler nation-states, and (2) mitigate the erasure and violence perpetrated by settler institutions of higher learning upon Indigenous communities and territories.

In this chapter, we introduce settler colonial theory and highlight key socio-ecological implications of settler colonialism that may be especially relevant to environmental sociologists. We elaborate on Kari Norgaard and James Fenelon's (2021) *Indigenous Environmental Sociology* to discuss how Indigenous scholars, communities, epistemologies, and methods are expanding sociological research and teaching, and steering environmental sociology toward a more interdisciplinary, intersectional, justice-oriented path. Finally, we discuss the ethical landscape of teaching and researching these topics from the confines of colonial institutions of higher learning, including land grant institutions directly implicated in Indigenous land dispossession. While the concepts discussed below may be relevant to Indigenous communities across the globe, our examples and cited literature primarily focus on a North American context given our community relations, collaborative histories, areas of scholarly focus, and responsibilities as academics at institutions situated in what today is known as the United States.

SETTLER COLONIAL THEORY PRIMER

Indigenous articulations of colonial systems have a long history. While we may never know the exact wording of some of the earliest critiques as they were orally delivered, there is also a long and ongoing record of written critiques from Indigenous peoples (some early examples include Apess; Eastman; Ša; Winnemuca; Standing Bear). Building on these Indigenous critiques and the work of Native scholars, settler colonial theory emerged to recognize the distinct socio-ecological processes that drive—and result from—settler colonialism. Whereas other forms of colonialism have as their primary motivation the extraction of natural resources and labor for the purposes of supplementing the metropole's economy, settler colonialism is a practice that aims to exert lasting and total control over a place through the installation of settlers. Scholars like Julie Matthews and Lucinda Aberdeen (2004), Evelyn Nakano Glenn (2015), Kyle P. Whyte (2018), and Patrick Wolfe (2006) theorize settler colonialism not as a one-time, long-ago event, but as an enduring social structure that continuously dispossesses Indigenous peoples of territory. Whyte (2018) describes it as an ongoing environmental injustice, "a social process by which at least one society seeks to establish its own collective continuance at the expense of the collective continuance of one or more other societies" (136). The imposition of settler colonial structures relies on two mutually reinforcing mechanisms: first, the removal of Indigenous peoples from the land via genocide, forced relocation, geographic containment, and assimilation; and second, the securing of territory for settlers by commodifying and privatizing land and instituting administrative and surveillance measures that can systematize and police its exclusive ownership (Fenelon and Trafzer 2014; Glenn 2015).

UNDERSTANDING INDIGENOUS SOVEREIGNTY AND RESISTANCE

Despite the brutality and erasure of settler colonial structures, Indigenous peoples persist and resist. J. Kēhaulani Kauanui (2016) clarifies that wherever settler colonialism exerts its force, Indigenous peoples subvert the system and keep Indigenous knowledge, practices, and ecological relationships in motion, a phenomenon Kauanui refers to as *enduring Indigeneity*.

Indigenous peoples are not passive victims of colonial violence but communities in active resistance exerting their power to counter, reshape, and decolonize colonial institutions (Hall and Fenelon 2015; Hormel and Norgaard 2009; Jacob 2013, 2016; LaDuke 2017; Simpson 2017; Steinman 2016; Whyte 2017).

Indigenous sovereignty can be conceptualized in various ways. A central understanding of Indigenous sovereignty is the political nationhood of Tribes with which the US government is legally mandated to uphold a trust responsibility and a government-to-government relationship. But beyond this state-centric definition are conceptualizations of sovereignty that encompass Indigenous peoples' freedom (as individuals or communities) to exercise their cultural practices and lifeways free from the violent constraints and assimilating forces of colonialism. Within this more expansive definition of sovereignty, Indigenous scholars and communities call for resistance to the oppressive colonial logics of patriarchy, heteronormativity, and capitalism (Alfred 2005; Coulthard 2014; Gilio-Whitaker 2019; Simpson 2017). Erich Steinman (2016) points to the need among sociologists to recognize Indigenous resistance beyond state-centered approaches. These tactics are vital, important, and worthy of sociological recognition.

Indigenous sovereignty in its various forms shapes Indigenous pursuits of environmental and social justice (Baldy 2013; Cantzler and Huynh 2016; Gilio-Whitaker 2019; Norgaard 2019; Steinman 2016; Yazzie and Baldy 2018). Dina Gilio-Whitaker (2019) outlines the ways in which mainstream environmental justice (EJ) frameworks neglect the distinct political, economic, and genocidal dimensions that shape environmental conditions for Indigenous peoples. She explains that for Indigenous peoples, a distributive EJ model (in which environmental benefits, resources, and burdens are equally distributed among all citizens regardless of identity) fails to recognize the history of Indigenous land dispossession and the desire of many Indigenous communities to resist capitalist incorporation. Of special significance to environmental sociologists is the strong relationship between Tribal sovereignty and ecological integrity. Steinman quotes Winona LaDuke (1999) to explain:

> A great variety of contemporary tribal actions can be understood not just as decolonizing efforts protecting the environment and natural resources but as protecting the material basis for the survival of distinctly Indigenous lifestyles, and upon that basis, Indigenous peoples. (LaDuke 1999; cited in Steinman 2016)

Indigenous identity and well-being are deeply dependent on ongoing, unencumbered relations with places, ecosystems, and species of cultural significance. As such, Indigenous sovereignty is deeply impacted by the imposition of settler ecologies at the expense of landscapes and species tended by Indigenous stewardship.

INTERROGATING THE NORMALIZATION OF SETTLER ECOLOGIES

Given the centrality of land in settler colonialism, scholars have emphasized the importance of ecology in understanding settler-Indigenous relations (Bacon 2018; Griffiths and Robin 1997; Whyte 2018). J.M. Bacon (2018) theorizes settler colonialism as an *eco-social structure* characterized by acts of *colonial ecological violence*. This violence takes many forms, ranging from the physical removal of Indigenous peoples via genocide and forced relocation to more subtle forms of "slow violence" that include the renaming of Indigenous places, the erasure of **Indigenous knowledges**, and the disruption of Indigenous ecologies via settler forms of land management.

As a philosopher, Kyle Whyte (2018) offers several conceptual framings that expand sociological theorizations of the environment. First, he equates ecology with "collective continuance," which he defines as "a community's capacity to be adaptive in ways sufficient for the livelihoods of its members to flourish into the future" (518). This definition of ecology interlocks human and more-than-human survival, wellbeing, and adaptive capacity. Furthermore, Whyte describes two ecological patterns of dispossession that help understand long-term impacts of settler colonialism: "vicious sedimentation" and "insidious loops." The term "vicious sedimentation" refers to "how constant ascriptions of settler ecologies onto Indigenous ecologies fortify settler ignorance against Indigenous peoples over time" (138). By altering landscapes managed by Indigenous peoples from time immemorial, settlers impose their own ecologies and erase evidence of Indigenous stewardship from the land, further reinforcing settler belonging. Insidious loops, on the other hand, refer to "the pattern of how historic settler industries that violated Indigenous peoples when they began are also implicated many years later in further environmental violence, such as climate injustice" (137–38). These concepts are highly useful in analyzing the ecological impacts of colonialism not as one-time events or linear progressions but as impacts that circle back and exponentially build on one another.

Both Whyte and Bacon mention the role of industry in settler colonial violence. Not surprisingly, settler ecologies are often rooted in capitalist production. The vicious sedimentation of settler and capitalist ecologies relies on Indigenous erasure and misrepresentation. Through these dual processes, most people living in settler states are kept unaware of many of the eco-social processes that define everyday interactions with the natural world. Contemporary battles over public education are one of the most obvious frontlines of this process, but when it comes to Indigenous issues, education has long been a site of manufactured ignorance on the part of non-Natives (Simpson 2014; Sabzalian 2019; Sabzalian et al. 2021). While today a few states mandate the inclusion of Native American history in public school education, the vast majority do not (Foxworth, Liu, and Sokhey 2015; Kruger 2019). Most people who have experienced education in the United States are woefully ignorant about Indigenous peoples while simultaneously imbibing a steady diet of stereotypes and misinformation regarding them (Kruger 2019; Davis-Delano et al. 2021). In this context, it is hardly surprising that the average US citizen has no clue who the original Indigenous peoples of their own hometowns are, how that land came to be under the authority of the settler state, or what treaty obligations are owed to original peoples. Importantly, settler narratives often hinge on a *terra nullius* conceptualization of North America that largely erases the role of Indigenous stewardship in producing abundant North American ecologies (Coulthard 2014).

ADDRESSING WESTERN SCIENTIFIC DOMINANCE AND THE ERASURE OF INDIGENOUS KNOWLEDGES

In the context of education and its role in erasing Indigenous peoples and ecologies, it is important to unpack the colonial role of Western science. Numerous scholars have examined Western science as a tool of colonialism (e.g., Dhillon 2020; Griffiths and Robin 1997; MacLeod 2000; Whit 2009). Western scientific pursuits have historically appropriated Indigenous knowledge for colonial gains, aligned scientific and colonial objectives, and misapplied Eurocentric Western science in non-European ecological and cultural contexts. Relatedly, Western science has also been utilized to construct racial boundaries and hierarchies (Dennis 1995; TallBear 2013), and to justify domination over women (Tuana 1989), sexual minorities (Foucault 1978), and other species (Haraway 1978).

Despite its claim to objectivity, Western science is neither value-free nor a-cultural (Haraway 1988; Levins and Lewontin 1985; Medin and Bang 2014). Scientists have social, cultural, and political identities that inevitably shape the research questions asked, the methods employed, how information is disseminated, and the ultimate objectives of scientific pursuits. Douglas Medin and Megan Bang (2014) point to the disproportionately white and male demographics of science, technology, engineering, and mathematics (STEM), which lead to Euro- and androcentric Western scientific practices. Even in a seemingly self-reflexive discipline like sociology, the settler lens manifests itself in the underrepresentation of Indigenous scholars and epistemologies, and the tendency to pathologize Indigenous peoples (Bacon 2017).

Importantly, the Western scientific enterprise (made up of scientists, research institutions, institutions of higher learning, and the entities that fund scientific pursuits) has historically dismissed traditional knowledge systems, including—especially—Indigenous knowledges (Medin and Bang 2014; Simpson 2004; Vinyeta 2022b). Tyler Jessen and colleagues (2021) explain that "[a] universal definition of [Indigenous knowledge] is precluded by its diversity. [Indigenous knowledge] is generally thought of as a body of place-based knowledges accumulated and transmitted across generations within specific cultural contexts" (93). Indigenous knowledges are central in maintaining Indigenous ecologies, and depend on Indigenous peoples' ability to maintain sovereign, reciprocal, multigenerational relations with culturally significant species and places.

Indigenous knowledges can be generalizable, but are often place-, culture-, and sometimes family- and gender-specific (Chisolm Hatfield et al. 2018; Jessen et al. 2021). This counters the emphasis in many Western academic disciplines to produce generalizable data with the broadest reach possible. If environmental sociology is to respond to the needs of Indigenous communities, to address colonial impacts on Indigenous ecologies, and to support Indigenous sovereignty, it must be willing to take a deep dive into the social dimensions of local ecologies and species assemblages. As Yvonne Sherwood (2016) explains in her description of Fourth World theory:

> Indigenous scholars voice concerns that decolonization is weakened by the practice of abstracting land into a "decolonial commons." As Schneider (2013) insists, current discussions of settler colonialism and the responses to the troubles it produces are undermined by treating "land as generic and equivalent." From this perspective, decolonial praxis must start with place. A Fourth World framework points out that the host world is neither abstract nor reproducible. (as cited in Sherwood 2016, 17–18)

For many environmental sociologists (especially settlers), moving beyond generic references to the "environment" and into the eco-cultural specificity of place will require interdisciplinary collaboration with ecologists and other natural scientists; more importantly, it will require consensual and ethical collaboration with Indigenous scientists and/or knowledge holders.

Finally, if environmental sociology is to honor Indigenous knowledges on Indigenous peoples' terms, then the subdiscipline must destigmatize the relationship between knowledge and spirituality that characterizes many Indigenous ways of knowing. Indigenous knowledge systems, spiritual practices, and ecosystem stewardship are often inseparable (Deloria and Wildcat 2001). Even in Indigenous communities where colonial religions have been adopted, there can be a melding of spirituality and Indigenous knowledges that imbues colonial religions with Indigenous values centering the environment and the power of the feminine (Jacob 2016). Many environmental sociologists rely strictly on Western scientific data to illustrate the environmental issues of importance to our field, assuming that other ways of knowing are reducible to unsubstantiated folklore. Yet there is a danger in such assumptions—Western scientists have much to learn from Indigenous knowledge holders, and as Kirsten Vinyeta's (2022b) analysis reveals, the consequences of erasing Indigenous knowledges can prove catastrophic for Indigenous peoples and ecosystems especially, but also for settlers. Even in less extreme cases, neglecting or erasing certain forms of knowledge means the subdiscipline only understands part of the picture, perpetuating a disconnect between how environmental problems and solutions are discussed in academic spaces, and how they are perceived and solved in actual communities.

EXPANDING WHO CONSTITUTES "THE SOCIAL"

In the last three decades, sociologists have increasingly incorporated more-than-human animals into sociological inquiry, with the notable rise of animal sociology (see Whitley and Vanselow, chapter 3 in this volume). These analyses vary in their approaches and theoretical orientations, from examining human and nonhuman play as a form of symbolic interactionism (Alger and Alger 1997; Jerolmack 2009; Sanders 2003), to utilizing Marxist frameworks to understand the alienation of dairy cows (Stuart, Schewe, and Gunderson 2013), to using political economic theory to understand why the advent of fossil fuels did not minimize whale hunting (York 2017). Yet important gaps remain. As Kari M. Norgaard and James Fenelon (2021) emphasize, Indigenous lifeways and knowledge systems—in which animals, plants, and even fire and water are often understood as agentic beings with whom Indigenous peoples hold reciprocal relationships—remain largely invisible. Settler framings of animal sociology obscure reciprocal relations that many Indigenous peoples and communities have with other species, specifically species living noncaptive lives. While there is danger in romanticizing Indigenous peoples' relationships with the environment, it is just as consequential to minimize the importance of ecological relations within Indigenous communities and therefore justify ongoing colonial violence (Ranco 2007).

Environmental and animal sociologists largely minimize the agency of other species and their role in shaping human social systems. However, there are some notable exceptions (e.g., Cherry 2019; Ergas and York 2021; Norgaard 2019; Pellow 2016; York and Longo 2017; York and Mancus 2013). For example, in theorizing critical environmental justice (CEJ) studies, David Pellow (2016) explains that "CEJ Studies recognizes that social inequality and oppression in all forms intersect and that members of the more-than-human world are subjects of oppression and agents of social change" (225). Richard York and colleagues describe dialectical relationships between humans and other species, in which other animals shape our lives as we shape theirs (York and Mancus 2013; York and Longo 2017). Christina Ergas and York (2021) make the case for sociological plant studies, recognizing plants as social agents deserving of long overdue attention within environmental sociology.

Indigenous scholars working largely outside sociology recognize the agency of other species and the vital importance of more-than-human relations in upholding Indigenous sovereignty (Coulthard 2014; Deloria and Wildcat 2001; Jacob 2016; Kimmerer 2013; LaDuke 2017; Lake et al. 2010; Salmón 2000; Simpson 2014; Whyte 2013, 2017; Wilson 2008 [to name a few]). Whyte (2017) uses the concept of *renewing relatives* to describe a process that involves both maintaining long-standing ecological relationships with species that are culturally vital, as well as forming new relationships with species that may be migrating because of climate change or other environmental factors. Glen Coulthard (2014) advances the concept of *grounded normativity* to describe "the modalities of Indigenous land-connected practices and longstanding experiential knowledge that inform and structure our ethical engagements with the world and our relationships with human and nonhuman others over time" (13). He explains that place-based, nondominant orientations toward land and more-than-humans defines Indigenous resistance to the intertwined structures of settler colonialism and capitalism. Environmental and animal sociologists might engage Indigenous conceptualizations of multispecies social systems to make space for healing and justice frameworks that dismantle Eurocentric nature-culture divides and instead encompass whole ecosystems.

REFRAMING VULNERABILITY AND RESILIENCE

For environmental sociologists focusing on disaster and climate change resilience, settler colonial theory is key to understanding the barriers that compromise Indigenous resilience. The 2022 IPCC report states with "high confidence" that colonialism is among the factors that *produce* climate vulnerability. Indigenous communities are not inherently vulnerable—they are historically adaptable communities with deep environmental knowledge that have been impacted by the multipronged impacts of colonialism that limit mobility, land management practices, intergenerational knowledge exchange, and resource access. And yet, against many odds, Indigenous communities are also at the forefront of some of the most profound and impactful climate adaptation and environmental restoration efforts worldwide. This dual reality must be honored by environmental sociologists if we are to heed Eve Tuck's (2009) call to "suspend damage-centered research" and instead center Indigenous desire as a guiding force for research. Viable solutions to the environmental crises of our time cannot be achieved if Indigenous peoples continue to be pathologized in the social sciences without a critical eye to the ongoing colonial violence that simultaneously compromises Indigenous resilience and the health of ecosystems worldwide.

Indigenous scholars within and beyond sociology are pointing to Indigenous knowledge systems, lifeways, and values as healing pathways towards the mitigation of socio-ecological damage inflicted by colonialism (Cochran et al. 2014; Fenelon and Alford 2020; Jacob et al. 2020; Lake and Christianson 2020; Martinez et al. 2023; Reo and Ogden 2018; Wildcat 2009 [to name a few]). Jacob and colleagues (2020) critique the lack of sociological engagement with "values emergent from the strengths and wisdom of Indigenous peoples," explaining that such values could really enhance scholarship focused on justice and equity. The authors introduce an *Indigenous values affirmation tool,* which includes ten vital Indigenous values—sense of humor, spirituality, being a community member, responsibility to relatives, relationship with the environment, respecting elders, self-determination, practicing my Indigenous culture and traditions, gratitude, and speaking one's Indigenous language. The authors explain that they:

> invite environmental sociologists, activists, and those working for justice in health, legal, education, and related academic spaces to engage Indigenous cultural values in efforts to challenge the exclusionary white spaces and counter the settler colonial violence that plagues all peoples. Doing so will allow for healing humans' relationships with the environment, our more than human relations, and with each other. (Jacob et al. 2020, 143)

Indigenous knowledges and cultural memories illuminate approaches to living that predate and can transcend neoliberal capitalist exploitation of people and land. Environmental sociology's overall lack of engagement with these knowledge and value systems may stem from scholars' tendencies to aspire to scholarship that is generalizable and applicable on a broad scale (in other words, more citable), neglecting the vital importance of localized efforts, knowledge, and resilience. However, as James Fenelon and Jennifer Alford (2020) assert, "widely implemented top-down centralized systems of decision-making about resource management, allocation and sustainability, dismiss the intrinsic value of including community centered knowledge in developing integrated models that meet the needs of both local and global communities" (373). If environmental sociology is to overturn its colonial undercurrents and embrace Indigenous epistemologies and methods, a turn towards the local—towards specific ecosystems, communities, and species assemblages—will be indispensable.

EXPANDING TEACHING AND METHODS IN SOCIOLOGY

Indigenous approaches to education can also inform our work as teachers of environmental sociology (Cajete 1994; Davidson and Davidson 2018; Deloria and Wildcat 2001; Jacob 2013; Jacob et al. 2018; Smith 2021; Simpson 2014). While most sociologists and students are not from Indigenous backgrounds or communities, embracing the values that are centered by many Indigenous educators can be transformative in the classroom. This is *not* a call to appropriate Indigenous culture, but rather an invitation to rethink how dominant pedagogical practices are not the best fit for promoting engaged and collaborative learning.

Consider how bizarre a standard exam might feel as a means of testing student knowledge after a term focused on Indigenous scholarship that stresses the vital importance of community, embodiment, and reciprocity. When we are including Indigenous research and creative materials in our classes it makes sense to also shift how we structure those classes. Some key concepts to consider come from Sara Florence Davidson and Robert Davidson (2018) in their book *Potlatch as Pedagogy*. While the book as a whole tells a culturally and even family-specific story of education and tradition, it also offers a set of principles for learning. We recommend that interested readers spend some time with this text. In the interest of brevity, consider this one principle: "Learning occurs through contribution." This will have a different meaning for different audiences of course, but in the context of the classroom it is worth asking ourselves the following questions: how do we create a course that allows students to learn in order to contribute? How do we create meaningful opportunities for them to learn in order to share? When it comes to environmental sociology, many of our students come looking for answers that we simply may not have. How can we invite them to come up with contributions that will get us closer to the goal of a more ecologically robust and socially just society?

When it comes to research, there are several crucial factors to consider. Tribes are sovereign nations with distinct knowledges, values, and rules of decorum, and must be active collaborators in, if not leading, research concerning their communities. Ideally, Indigenous scholars or knowledge holders from the community in question can lead research efforts. This requires institutions of higher learning to recognize Indigenous knowledges as valuable and Indigenous students, faculty, and staff as indispensable and worthy of institutional transformation. Settlers, too, can collaborate with Indigenous communities, but there is much to be learned before this can ethically take place. Understanding the principles of decolonial and Indigenous-led approaches to research is imperative. While we must interrogate whether research taking place at colonial institutions can ever be truly decolonial, the protocols introduced in works such as Linda Tuhiwai Smith's (2021) *Decolonizing Methodologies: Research and Indigenous Peoples* and Shawn Wilson's (2019) *Research Is Ceremony: Indigenous Research Methods* can illuminate how one might enter into ethical and meaningful research relations that recognize the cultural and political dimensions of work with Indigenous communities. It is important to note that every Tribe has different approaches to research and collaboration. Some Tribes may not be interested in collaborating with academic researchers, whereas other Tribes have long-term histories of academic collaboration. There are Tribes that have their own institutional review board (IRB) processes that must be approved by Tribal review boards before research can commence (such as the Navajo Nation and the Karuk Tribe, for example). These Tribal IRB requirements are separate from academic IRB processes, but just as—if not more—important in protecting Tribal knowledge, interests, landscapes, and peoples.

Research methods, too, must be considered when engaging in collaborative research with Indigenous communities. Indigenous collaborators and participants may prefer oral and visual communication over written exchanges and communal research processes over individual ones (Wilson 2019). As such, methods involving storytelling (see Archibald 2008 for her *Storywork* methodology), visual media (Cordes 2021; Vinyeta 2022a) and conversational focus groups may be preferable to surveys and individual interviews. Perhaps of utmost importance is the protection of knowledge and data sovereignty, as well as the establishment of clear protocols to ensure Tribal ownership over data and the protection of culturally sensitive information, such as the location and nature of sacred sites. There is a rapidly emerging literature introducing guiding principles for Indigenous data governance and knowledge sovereignty that can guide environmental sociologists working collaboratively with Indigenous communities (e.g., Jennings et al. 2023; Kukutai and Taylor 2016; Latulippe and Klenk 2020).

DISMANTLING SETTLER COLONIAL VIOLENCE IN THE ACADEMY AND BEYOND

In recent years, there has been a growing conversation about "decolonizing" the academy. One common practice that has gained traction—first in Canada and more recently in the United States is the process of land acknowledgements. While land acknowledgements can be part of a meaningful protocol of respect between Indigenous guests and Indigenous hosts, when formalized by non-Native institutions they may take on a different meaning, one that is largely superficial (Dua and Coburn forthcoming). Consider, for example, the critiques leveled by Hayden King (2019), which suggest that "There is a danger of the acknowledgement just becoming that excuse, through, which these institutions provide themselves permission to be on that territory." Furthermore, Theresa Ambo and Theresa Rocha Beardall's (2023) systematic analysis of land acknowledgement practices at land grant (or, as the authors refer to them, *land grab*) universities reveals the *rhetorical removal* of Indigenous peoples by speaking of Indigenous peoples as relics of the past with little to no contemporary agency. Both King (2019) and Ambo and Rocha Beardall (2023) point to the importance of matching institutional action with land acknowledgements.

This focus on the more-than-symbolic is central to any effort to transform oppressive relations. In their well-cited work "Decolonization Is Not a Metaphor," Eve Tuck and K. Wayne Yang (2012) impress on readers the need for discussions of decolonization to center Indigenous peoples, sovereignty, and the return of land. In this light, the question must be asked: Can the academy even be decolonized? That's a question many are wrestling with, but one thing is clear: no effort toward decolonization can be meaningful without addressing issues of land and power. For environmental sociologists, the central questions of our field are intimately related to the central questions of decolonization, and we would be remiss to ignore that. Furthermore, environmental sociologists—especially those employed at land grant institutions—can play vital roles in driving reparations, collaboration, and co-management with Tribes that have been dispossessed by their universities.

Engaging with the way colonialism has informed and continues to shape eco-social relations is pivotal for environmental sociology. So too is meaningful and respectful engagement with Indigenous peoples, knowledges, and practices. This is no simple task. As Taiaiake Alfred (2004) reminds us:

> Colonialism is a total relation of power, and it has shaped the existence not only of those who have suffered its effects but also those who have profited from it. The culture we have inherited is thoroughly infused with the values of domination and submission, fear and compliance, and the act of unrestrained and unthinking consumption that is the engine of our economic and political system. (91)

Within this total relation of power, the academy is not a place apart but rather a place that is deeply implicated and thoroughly imbued with the norms and goals of colonial relations. As environmental sociologists, it is our job to reveal the power relations that shape our eco-social reality. If we are truly invested in changing those power relations, then change must begin with us and include how we approach our research, teaching, and our relationships with others including both the human and the more-than-human.

REFERENCES

Alfred, Taiaiake. 2004. "Warrior Scholarship: Seeing the University as a Ground of Contention." In *Indigenizing the Academy: Transforming Scholarship and Empowering,* edited by Wilson, Angela C., and Devon A. Mihesuah. 2004. University of Nebraska Press.

Alfred, Taiaiake. 2005. *Wasa'se : Indigenous Pathways of Action and Freedom.* Broadview Press.

Alger, Janet M., and Steven F. Alger. 1997. "Beyond Mead: Symbolic Interaction between Humans and Felines." *Society & Animals* 5 (1): 65–81.

Ambo, Theresa, and Theresa Rocha Beardall. 2023. "Performance or Progress? The Physical and Rhetorical Removal of Indigenous Peoples in Settler Land Acknowledgments at Land-Grab Universities." *American Educational Research Journal* 60 (1): 103–40.

Archibald, Jo-ann (Q'um Q'um Xiiem). 2008. *Indigenous Storywork: Educating the Heart, Mind, Body, and Spirit.* University of British Columbia Press.

Bacon, J.M. 2017. "At Risk-ing and Asterisk-ing Native Peoples in Social Science Research: The Invisibilizing of Settler-Colonialism Made Visible Through Keyword Network Analysis." Paper presented at the Annual Meeting of the American Sociological Association. Montréal, Quebec, Canada.

Bacon, J.M. 2018. "Settler Colonialism as Eco-Social Structure and the Production Of Colonial Ecological Violence." *Environmental Sociology* 5 (1): 59–69.

Baldy, Cutcha Risling. 2013. "Why We Gather: Traditional Gathering in Native Northwest California and the Future of Bio-Cultural Sovereignty." *Ecological Processes* 2 (1): 1–10.

Cantzler, Julia Miller, and Megan Huynh. 2016. "Native American Environmental Justice as Decolonization." *American Behavioral Scientist* 60 (2): 203–23.

Carter, Bob, and Nickie Charles. 2018. "The Animal Challenge to Sociology." *European Journal of Social Theory* 21 (1): 79–97.

Cajete, Gregory. 1994. *Look to the Mountain: An Ecology of Indigenous Education*. Kivakí Press.

Chisholm Hatfield, Samantha, Elizabeth Marino, Kyle Powys Whyte, Kathie D. Dello. and Philip W. Mote. 2018. "Indian Time: Time, Seasonality, and Culture in Traditional Ecological Knowledge of Climate Change." *Ecological Processes* 7 (1): 1–11.

Cherry, Elizabeth. 2019. *For the Birds: Protecting Wildlife through the Naturalist Gaze*. Rutgers University Press.

Cochran, Patricia, Orville H. Huntington, Caleb Pungowiyi, Stanley Tom, F. Stuart Chapin III, Henry P. Huntington, Nancy G. Maynard, and Sarah F. Trainor. 2013. "Indigenous Frameworks for Observing and Responding to Climate Change in Alaska." *Climatic Change* 120 (3): 557–67.

Collins, Patricia Hill. 2022. *Black Feminist Thought: Knowledge, Consciousness, and the Politics of Empowerment*. Routledge.

Cordes, Ashley. 2021. "Revisiting Stories and Voices of the Rogue River War (1853–1856): A Digital Constellatory Autoethnographic Mode of Indigenous Archaeology." *Cultural Studies ↔ Critical Methodologies* 21 (1): 56–69.

Coulthard, Glen Sean. 2014. *Red Skin, White Masks: Rejecting the Colonial Politics of Recognition*. University of Minnesota Press.

Davidson, Sara Florence, and Robert Davidson. 2018. *Potlatch as Pedagogy: Learning through Ceremony*. Portage & Main Press.

Davis-Delano, Laurel R., Jennifer J. Folsom, Virginia McLaurin, Arianne E. Eason, and Stephanie A. Fryberg. 2021. "Representations of Native Americans in U.S. Culture? A Case of Omissions and Commissions." *Social Science Journal*, September 15, 2021..

Deloria, Vine, and Daniel R. Wildcat, Jr. 2001. *Power and Place: Indian Education in America*. American Indian Graduate Center and Fulcrum Resources.

Dennis, Rutledge M. 1995. "Social Darwinism, Scientific Racism, and the Metaphysics of Race." *Journal of Negro Education* 64 (3): 243–52.

Dhillon, Carla M. 2020. "Indigenous Feminisms: Disturbing Colonialism in Environmental Science Partnerships." *Sociology of Race and Ethnicity* 6 (4): 483–500.

Dua, Enakshi, and Elaine Coburn. Forthcoming. "Land Acknowledgements in an Era of Reconciliation." In *Settler Colonialism in Canada: Perspectives,*

Comparisons, Cases. edited by D. MacDonald and E. Grafton. Vol. 1. University of Regina Press.
Ergas, Christina, and Richard York. 2023. "A Plant by any Other Name: . . . Foundations for Materialist Sociological Plant Studies." *Journal of Sociology* 59 (1): 3–19.
Fenelon, James, and Jennifer Alford. 2020. "Envisioning Indigenous Models for Social and Ecological Change in the Anthropocene." *Journal of World-Systems Research* 26 (2): 372–99.
Fenelon, James V., and Clifford E. Trafzer. 2014. "From Colonialism to Denial of California Genocide to Misrepresentations: Special Issue on Indigenous Struggles in the Americas." *American Behavioral Scientist* 58 (1): 3–29.
Ferrara, Enzo. 2015. "Who's Asking? Native Science, Western Science and Science Education." *Electronic Green Journal* 38 (Spring).
Foucault, Michel. 2012. *The History of Sexuality*. Vintage.
Foxworth, Raymond, Amy H. Liu, and Anand Edward Sokhey. 2015. "Incorporating Native American History into the Curriculum: Descriptive Representation Or Campaign Contributions?: Incorporating Native American History into the Curriculum." *Social Science Quarterly* 96 (4): 955–69.
Gilio-Whitaker, Dina. 2019. *As Long as Grass Grows: The Indigenous Fight for Environmental Justice, from Colonization to Standing Rock.* Beacon Press.
Glenn, E.N. 2015. "Settler Colonialism as Structure: A Framework for Comparative Studies of U.S. Race and Gender Formation." *Sociology of Race and Ethnicity*, *1*(1), 52–72
Griffiths, Tom, and Libby Robin. 1997. *Ecology and Empire: Environmental History of Settler Societies.* Edinburgh University Press.
Hall, Thomas D., and James V. Fenelon. 2015. *Indigenous Peoples and Globalization: Resistance and Revitalization.* Paradigm.
Haraway, Donna. 1988. "Situated Knowledges: The Science Question in Feminism and the Privilege of Partial Perspective." *Feminist Studies* 14 (3): 575–99.
Hormel, Leontina M., and Kari M. Norgaard. 2009. "Bring the Salmon Home! Karuk Challenges to Capitalist Incorporation." *Critical Sociology* 35 (3): 343–66.
Jacob, Michelle M. 2013. *Yakama Rising: Indigenous Cultural Revitalization, Activism, and Healing.* University of Arizona Press.
Jacob, Michelle M. 2016. *Indian Pilgrims: Indigenous Journeys of Activism and Healing with Saint Kateri Tekakwitha.* University of Arizona Press.
Jacob, Michelle M., Leilani Sabzalian, Joana Jansen, Tary J. Tobin, Claudia G. Vincent, and Kelly M. LaChance. 2018. "The Gift of Education: How Indigenous Knowledges Can Transform the Future of Public Education." *International Journal of Multicultural Education* 20 (1): 157–85.
Jacob, Michelle M., Kelly L. Gonzales, Deanna Chappell Belcher, Jennifer L. Ruef, and Stephany RunningHawk Johnson. 2020. "Indigenous Cultural Values Counter the Damages of White Settler Colonialism." *Environmental Sociology* 7 (2): 134–46.

Jennings, Lydia, Talia Anderson, Andrew Martinez, Rogena Sterling, Dominique David Chavez, Ibrahim Garba, Maui Hudson, Nanibaa' A. Garrison, and Stephanie Russo Carroll. 2023. "Applying the 'CARE Principles for Indigenous Data Governance' to Ecology and Biodiversity Research." *Nature Ecology & Evolution* 7:1547–1551.

Jerolmack, Colin. 2009. "Humans, Animals, and Play: Theorizing Interaction When Intersubjectivity Is Problematic." *Sociological Theory* 27 (4): 371–89.

Jessen, Tyler D., Natalie C. Ban, Nicholas XEM OLTW Claxton, and Chris T. Darimont. 2022. "Contributions of Indigenous Knowledge to Ecological and Evolutionary Understanding." *Frontiers in Ecology and the Environment* 20 (2): 93–101.

Kauanui, J. Kēhaulani. 2016. "A Structure, Not an Event: Settler Colonialism and Enduring Indigeneity." *Lateral* 5 (1).

King, Hayden. 2019. " 'I Regret It:' Hayden King on Writing Ryerson University's Territorial Acknowledgement." CBC Radio. January 18. https://www.cbc.ca/radio/unreserved/redrawing-the-lines-1.4973363/i-regret-it-hayden-king-on-writing-ryerson-university-s-territorial-acknowledgement-1.4973371.

Krueger, Justin. 2019. "To Challenge the Settler Colonial Narrative of Native Americans in Social Studies Curriculum: A New Way Forward for Teachers." *History* Teacher 52 (2): 291–318.

Kukutai, Tahu, and John Taylor, eds. 2016. *Indigenous Data Sovereignty: Toward an Agenda*. Australian National University Press.

LaDuke, Winona. 2017. *All Our Relations: Native Struggles for Land and Life*. Haymarket Books.

Lake, Frank K., and Amy Cardinal Christianson. 2020. "Indigenous fire stewardship." In *Encyclopedia of Wildfires and Wildland-Urban Interface (WUI) Fires*. Springer.

Latulippe, Nicole, and Nicole Klenk. 2020. "Making Room and Moving Over: Knowledge Co-Production, Indigenous Knowledge Sovereignty and the Politics of Global Environmental Change Decision-Making." *Current Opinion in Environmental Sustainability* 42:7–14.

Levins, Richard, and Richard Lewontin. 1985. *The Dialectical Biologist*. Harvard University Press.

MacLeod, Roy. 2000. *Nature and Empire: Science and the Colonial Enterprise*. University of Chicago Press.

Martinez, Denniss J., Clare E. B. Cannon, Alex McInturff, Peter S. Alagona, and David N. Pellow. 2023. "Back to the Future: Indigenous Relationality, Kincentricity and the North American Model of Wildlife Management." *Environmental Science & Policy* 140:202–7.

Matthews, Julie, and Lucinda Aberdeen. 2004. "Racism, Racialisation and Settler Colonialism." In *Disrupting Preconceptions: Postcolonialism and Education*, edited by A. Hickling-Hudson, J. Matthews, and A. Woods. Post Pressed.

Norgaard, Kari M. 2019. *Salmon and Acorns Feed Our People: Colonialism, Nature, and Social Action*. Rutgers University Press.

Norgaard, Kari M., Fenelon, James V. 2021. "Towards an Indigenous Environmental Sociology." In *Towards an Indigenous Environmental Sociology*. Springer.

Pellow, David N. 2016. "Toward a Critical Environmental Justice: Black Lives Matter as an Environmental Justice Challenge." *Du Bois Review* 13 (2): 221–36.

Ranco, Darren J. 2007. "The Ecological Indian and the Politics of Representation." In *Native Americans and the Environment: Perspectives on the Ecological Indian*, edited by M. E. Harkin and D. R. Lewis. University of Nebraska Press.

Reo, Nicholas J., and Laura A. Ogden. 2018. "Anishnaabe Aki: An Indigenous Perspective on the Global Threat of Invasive Species." *Sustainability Science* 13:1443–52.

Rocha Beardall, Theresa. 2022. "Settler Simultaneity and Anti-Indigenous Racism at Land-Grant Universities." *Sociology of Race and Ethnicity* 8 (1): 197–212.

Sabzalian, Leilani. 2019. "The Tensions between Indigenous Sovereignty and Multicultural Citizenship Education: Toward an Anticolonial Approach to Civic Education." *Theory & Research in Social Education* 47 (3): 311–46.

Sabzalian, Leilani, Sarah B. Shear, and Jimmy Snyder. 2021. "Standardizing Indigenous Erasure: A TribalCrit and QuantCrit Analysis of K–12 US Civics and Government Standards." *Theory & Research in Social Education* 49 (3): 321–59.

Sanders, Clinton R. 2003. "Actions Speak Louder than Words: Close Relationships between Humans and Nonhuman Animals." *Symbolic Interaction* 26 (3): 405–26.

Sherwood, Yvonne P. 2016. "Toward, with and from a Fourth World." *Fourth World Journal* 14 (2): 15–26.

Simpson, Leanne R. 2004. "Anticolonial Strategies for the Recovery and Maintenance of Indigenous Knowledge." *American Indian Quarterly* 28 (3): 373–84.

Simpson, Leanne Betasamosake. 2014. "Land as Pedagogy: Nishnaabeg Intelligence and Rebellious Transformation." *Decolonization: Indigeneity, Education & Society* 3 (3).

Simpson, Leanne Betasamosake. 2017. *As We Have Always Done: Indigenous Freedom through Radical Resistance*. University of Minnesota Press.

Smith, Linda Tuhiwai. 2021. *Decolonizing Methodologies: Research and Indigenous Peoples*. Bloomsbury.

Steinman, Erich W. 2016. "Decolonization Not Inclusion: Indigenous Resistance to American Settler Colonialism." *Sociology of Race and Ethnicity* 2 (2): 219–36.

Stuart, Diana, Rebecca L. Schewe, and Ryan Gunderson. 2013. "Extending Social Theory to Farm Animals: Addressing Alienation in the Dairy Sector." *Sociologia Ruralis* 53 (2): 201–22.

TallBear, Kim. 2013. *Native American DNA : Tribal Belonging and the False Promise of Genetic Science*. University of Minnesota Press.

Tovey, Hilary. 2003. "Theorising Nature and Society in Sociology: The Invisibility of Animals." *Sociologia Ruralis* 43 (3): 196–215.

Tuana, Nancy, ed. 1989. *Feminism and Science.* Indiana University Press.
Tuck, Eve. 2009. "Suspending Damage: A Letter to Communities." *Harvard Educational Review* 79 (3): 409–28.
Tuck, Eve, and K. Wayne Yang. 2012. "Decolonization Is Not a Metaphor." *Education & Society* 1 (1): 1–40.
Tuck, Eve, and Marcia McKenzie. 2015. *Place in Research: Theory, Methodology, and Methods.* Routledge.
Veracini, Lorenzo. 2011. "Introducing: Settler Colonial Studies." *Settler Colonial Studies* 1 (1): 1–12.
Vinyeta, Kirsten. 2022a. *Ikpíkyav (To Fix Again): Drawing from Karuk World Renewal to Contest Settler Discourses of Vulnerability.* PhD diss., University of Oregon.
Vinyeta, Kirsten. 2022b. "Under the Guise of Science: How the US Forest Service Deployed Settler Colonial and Racist Logics to Advance an Unsubstantiated Fire Suppression Agenda." *Environmental Sociology* 8 (2): 134–48.
Whit, Laurelyn. 2009. *Science, Colonialism and Indigenous Peoples: The Cultural Politics of Law and Knowledge.* Cambridge University Press.
Whyte, Kyle Powys. 2013. "On the Role of Traditional Ecological Knowledge as a Collaborative Concept: A Philosophical Study." *Ecological Processes* 2 (1): 1–12.
Whyte, Kyle Powys. 2017. "Indigenous Climate Change Studies: Indigenizing Futures, Decolonizing the Anthropocene." *English Language Notes* 55 (1–2): 153–62.
Whyte, Kyle Powys. 2018. "Settler Colonialism, Ecology, and Environmental Injustice." *Environment and Society* 9 (1): 125–44.
Wildcat, Daniel R. 2009. *Red Alert! Saving the Planet with Indigenous Knowledge.* Fulcrum.
Wilson, Shawn. 2019. *Research Is Ceremony: Indigenous Research Methods.* Fernwood.
Wolfe, Patrick. 2006. "Settler Colonialism and the Elimination of the Native." *Journal of Genocide Research* 8 (4): 387–409.
Yazzie, Melanie, and Cutcha Risling Baldy. 2018. "Introduction: Indigenous peoples and the politics of water." *Decolonization: Indigeneity, Education & Society* 7 (1): 1–18.
York, Richard. 2017. "Why Petroleum did Not Save the Whales." *Socius: Sociological Research for a Dynamic World,* 3:1–13.
York, Richard, and Stefano B. Longo. 2017. "Animals in the World: A Materialist Approach to Sociological Animal Studies." *Journal of Sociology* 53 (1): 32–46.
York, Richard, and Philip Mancus. 2013. "The Invisible Animal: Anthrozoology and Macrosociology." *Sociological Theory* 31 (1): 75–91.

7 An Anticolonial Approach to Environmental Sociology

Michael Warren Murphy

For several decades, environmental sociologists have stretched the sociological imagination beyond the traditional limits of "humanity" and "society." Close to forty years ago, William Catton and Riley Dunlap (1978) argued that sociology had reached an impasse whereby the discipline was ill equipped to handle the challenges posed by the rise of environmental problems and their destabilizing effect on "American dreams of social progress, upward mobility, and societal stability" (42). A new environmental or ecological paradigm was implicit in the emergence of environmental sociology, one that challenged the anthropocentric inclinations of the mainstream "human exceptionalism paradigm" and its tendency to bifurcate society and nature.

Whereas environmental sociology emerged against the bifurcating tendencies of an anthropocentric worldview, proponents of decolonial or postcolonial sociology point to a different form of bifurcation, one characteristic of sociology's Eurocentrism. As Julian Go (2013) argues, "Sociology's Orientalism, Eurocentrism and suppression of imperial history are problematic not just in themselves but also because they lead us to analytically bifurcate connections and thus overlook the real social relations by which sociologists' main object—"modern society" and its boundaries—has been constituted" (38).

Taken together, these sociological interventions point to two analytic bifurcations at the core of the discipline. On the one hand, sociology has neglected to take seriously its own imperial origins and the ways in which its categories of thought and modes of analysis are shaped by it. On the other hand, the discipline has maintained an anthropocentric worldview that neglects the biophysical dimensions that make social life possible in the first place. In anticipation of the work ahead for the next several decades, over which we are likely to see and/or experience the intensification of the socioecological disaster that has been unfolding for the past five hundred years of modern racial-colonial capitalism (see chapter 5 on racial capitalism), this chapter points to the importance of an anticolonial approach to environmental sociological inquiry.

PIONEERING ENVIRONMENTAL SOCIOLOGY

> The seeds of great hopes were spread by the Europeans over their dominions. They were the daydreams of an expanding culture that promised instant fruition, if not today, then tomorrow, and if not in the Old World, then in the New. History has not been kind to these aspirations; its cruel realities are descending upon a people unaccustomed to the edge of frustration so that daydreams seem transformed into nightmares.
>
> WILLIAM R. BURCH JR.

Written by William R. Burch Jr., a sociologist employed by the Yale School of Forestry, *Daydreams and Nightmares: A Sociological Essay on the American Environment* (1971) represents an important precursor in the pioneering of environmental sociology. In this book, Burch argued that sociologists should extend their disciplinary focus to include the earth's air, water, soil, mineral, plant, and animal participants in life by examining the ways that myth, social structure, and ecosystem interpenetrate. Like the pioneers that would follow, Burch was motivated by a need to retool sociology to confront a frightening new reality that emerged from the thwarted aspirations and dashed hopes of Euro-Western civilization in its discovery of the environmental crisis. His book was an attempt to slow down and think carefully about how American environmental problems developed in an ethos permeated by "endless crusades of nervous action" to address them (8).

In his review of *Daydreams and Nightmares*, William Catton (1972a) emphasized the shifting social and ecological terrain on which the book was staked. In Catton's estimation, the book was a product of things happening to human society and the field of sociology at a time when "the traditional American expectation of a future always brighter than the past [had] become inverted," and sociology was in disarray "partly from the difficulties of adjusting to this upheaval in [the] intellectual climate" (240). Foreshadowing his collaborative articulation of environmental sociology with Riley Dunlap several years later, Catton thought that Burch's book made it "abundantly clear that sociologists [could] no longer afford to ignore the fact that ecological changes significantly influence human social systems" (240). Catton applauded Burch's refusal of political scapegoating and "ideological finger wagging" (Burch 1971, 39). Rather than viewing arrogance and racism as a uniquely American flaw, Catton (1972a) maintained that they were the "tragic by-products of the invasion and exploitation of a new habitat by an exuberant and culturally pre-adapted population" (241). He not only understated the environmental significance of the imperial invasion and racial-colonial exploitation of the land that would become America's, but he also echoed the anti-political normativity that represented the sort of environmental-sociological inquiry he valued most (Catton 1972b).

As the most recognized pioneers of environmental sociology, William R. Catton Jr. and Riley E. Dunlap were pivotal in staking claims of new intellectual territory by defining and legitimating this emergent field of thought. Catton and Dunlap argued that the emergence of environmental sociology signaled a paradigmatic shift from the old assumptions of sociology to new ones. The former sociological paradigm, referred to as the "human exceptionalism paradigm," was not only anthropocentric but also characterized by a set of assumptions that placed humans and human society above and beyond biophysical limits. With growing concern about environmental problems and crises, sociologists began to rethink these assumptions, leading to the emergence of environmental sociology, which brought with it a new set of assumptions under a "new ecological paradigm"—namely, that human beings are one of many interdependent species in a broader biotic community, that cause and effect in this web of nature can produce unintended consequences from human action, and that the world is finite with physical and biological limits constraining economic growth and social progress (Catton and Dunlap 1978). Thus, Catton and Dunlap defined environmental sociology as "the study of interaction between the environment and society," which "involves studying the effects of the environment on society (e.g., resource abundance or scarcity on stratification) and the effects of society on the environment (e.g., the contributions of differing economic systems to environmental degradation)" (44).

Despite articulating what William Catton and Riley Dunlap understood as a paradigmatic shift, we must ask the following question: is it conceivable that while specifying and legitimating this new domain of sociological inquiry, these pioneers carried forward certain metatheoretical (ontological, epistemological, and axiological) tendencies and assumptions?

In terms of ontology, or the domain of knowledge concerned with existence or being—and leaving aside the society-environment dualism that others would later contend with (see, e.g., Carolan 2005, Freudenburg,

Frickel, and Gramling 1995, Rice 2013, White, Rudy, and Gareau 2016)—the pioneers were concerned foremost with the existence of environmental problems, in addition to the environmental variables that mainstream sociology tended to ignore. For instance, Catton was particularly attuned to "congestion" in the parks and campgrounds that his family frequented for recreation. In fact, his concerns about population growth in the Puget Sound region led him to resign from the University of Washington's sociology faculty and move his family to Christchurch, New Zealand, another settler-colonized region of the earth, whose environment he noted resembled "an earlier (less populous) version of western Washington" (Catton 2008, 473). Ecological factors like population pressure mattered insofar as they pointed to the *reality* of biophysical limits to social progress and economic growth.

Speaking of reality, Catton and Dunlap's environmental sociology was premised on a specific variation of realist epistemology in which what is known about the world exists independently of our conceptualization of it. As a self-identified neopositivist, Catton had a "special desire to contribute to pushing sociology toward becoming more truly scientific" (472). While living in New Zealand, he discovered the book *Violence, Monkeys, and Man* by Claire and W. M. S. Russell, which, along with his "aversion to the local impacts of population pressure in the Pacific Northwest," reinforced his decision to read more work by biologists and ecologists, as well as "to pay more attention to some of the biogeochemical processes and other factors traditionally deemed nonsocial and 'irrelevant' to sociology" (473). The new ecological paradigm of sociology made it possible to account for that which sociology had long ignored, and this is what the pioneers argued secured its relevance beyond science.

Thus, regarding axiological considerations, or the normative aspects of intellectual life, environmental sociology was of value because it could guide human societies to a better, more sustainable future. As Catton put it, "Without the [new ecological paradigm], sociologists would provide inadequate guidance to decision makers and our fellow citizens" (476). How exactly would environmental sociologists play such a vital role? By considering the "social organizational requirements of such a [sustainable] society—ranging from energy efficient housing patterns to zero population growth—but they must also ask how existing societies might be changed to meet such requirements" (Dunlap and Catton 1979, 266).

Although environmental sociology represented a departure from the status quo, these overlapping metatheoretical assumptions illustrated the extent to which the field's original concerns, objects of study, conceptual tools, and theoretical framing remained lodged within a particular perspective on the world that is still too often taken for granted. From whose perspective is a statement like the following written? "We and our immediate ancestors lived through an age of exuberant growth, overshooting permanent carrying capacity without knowing what we were doing" (Catton 1980, 5). From whose perspective could the past several centuries have been considered a period of "magnificent progress," given the attendant racial-colonial violence, as well as the displacement, eradication, and enslavement of countless people and other life-forms?[1] Is there a standpoint from which it is clear that the *seeds of great hope that Europeans planted over their dominions* had always yielded terror, such that their daydreams were never anything but a tremendous nightmare?

EMPIRE, COLONIALISM, AND SOCIOLOGICAL THOUGHT

Distinguished by multiple tactics, practices, and modes of power, including both direct or indirect rule, **empires** are "hierarchically ordered formations wherein a state or center exercises control or unequal influence over subordinated territories, peoples, and societies" (Go 2011, 12). Although **imperialism** and **colonialism** are closely related social forms, and are often treated as synonymous, imperialism, at least as George Steinmetz (2014) defines it, is a strategy of political rule over foreign peoples and lands that does not always involve conquest, whereas colonialism entails direct violence and the coercion of subjugated populations (79). As with empire, modern colonialism is characterized by a multiplicity of nested and fused forms—such as penal colonies and militarized zones, plantation colonies, settler colonies, protectorate colonies, and port colonies (Manjapra 2020).

Five hundred or so years of colonialism and European empire have profoundly shaped the modern world, along with how we know or understand it. Given the longevity of imperialism as a political form throughout social history—with the earliest known instance taking shape in Mesopotamia around 2,350 BCE with the Akkadian Empire (Foster 2015)—it is important to ask what is different about the modern situation? In his examination of colonialism in a global perspective, Kris Manjapra (2020) argues that it is distinct from imperialism as well as premodern forms of colonialism in the sense that it accompanied the rise of capitalist empires beginning in the 1400s (CE) onward:

> The new colonialism—that is, colonial force arising through the history of racial capitalism—was more violent and more invasive than any other form of colonialism that the world had previously known. Vast

> amounts of Native lands were confiscated through genocidal wars. More African people were kidnapped, trafficked, and incarcerated under slavery than ever before or after (more than sixty million people over the course of five hundred years). Almost 90 percent of all peasants and rural people worldwide eventually had their lands expropriated from them. Almost 98 percent of the globe's territory gradually experienced some degree of long-term colonial occupation. Wealth disparities developed that eventually allotted 10 percent of people on earth (mostly living in Europe, North America, and other European settler states) ownership of 85 percent of the world's wealth. Centuries of colonialism permanently changed life and economies on a worldwide scale. (8–9)

The global hierarchy created by the machinery of modern empire and colonialism is reflected in both the material inequalities and epistemic barriers between the so-called Global North and South (Al-Hardan 2022; Connell 2007).

As an intellectual provocation and practice, postcolonial and decolonial sociology rests on normative (or axiological) concerns with countering colonial domination through subversive social knowledge. In other words, though the underlying genealogies of postcolonial and decolonial thought are distinct, they both turn on normative judgements about the values of inquiry, and what is valued is knowledge that goes against the grain (Bhambra 2014b). For this reason, and to avoid the temporal confusions that emerge from the use of the prefix "post" in postcolonial, I refer to anticolonial sociology as any approach that shares this fundamental axiological assumption. That said, the more important question is, what does this look like in practice? To address this, I want to point to Julian Go's (2016) intervention, while recognizing that there are other relevant engagements worthy of attention (e.g., Bhambra 2014a; Bhambra and Holmwood 2021; Meghji 2021).

In *Postcolonial Thought and Social Theory*, Go suggests two key analytic moves in contradistinction to what constitutes most projects in mainstream sociology. First, he argues, we must rely on what he calls postcolonial relationism. Given sociology's tendency toward analytically bifurcating the world, thereby obfuscating how colonizer/colonized, metropole/colony, and nature/society are mutually constituted, Go suggests that we replace "the imperial episteme's law of division with a methodological law of connection: sustained examinations of mutual connection across expansive social space" (114). Postcolonial relationism "attends to the mutual constitution of the powerful and powerless, the metropole and colony, the core and the postcolony, the Global North and Global South . . . taken up to the geopolitical scene, scaled upward and outward to critically apprehend imperial interaction and their enduring legacies" (142). Therefore, for example, rather than talking about capitalism alone and starting with the European experience of the white male worker in the factories of European societies, we might look to the ways in which capitalism is built on and supported by the racial-colonial practices of social domination and exploitation. We cannot think about capitalist development in Europe and North America, therefore, without reference to the global imperial connections that made it possible (Bhambra 2020).

The second analytic move is to proceed from the subaltern standpoint, rooted in perspectival realism, which aims to overcome sociology's Eurocentrism and false universalism. Go argues that perspectival realism "insists that there is a real world with observable and knowable features (realism) but that what we see in that world, how we describe it, and what we think about it partially depends upon the observer and his or her means of observation (constructivism)" (163). In other words, where an observer stands in the social world shapes what they can or cannot see. The subaltern standpoint, then, refers to the social location and perspective of colonially subjugated groups. In the same way that the standpoint of women in society is necessary to better apprehend the nature of patriarchy, we must rely on the subaltern standpoint to understand the enduring impact of colonialism. How do we work with subaltern standpoint(s)? Go suggests that we must "suspend or circumvent the analytic categories constructed from the metropolitan-imperial standpoint and instead start from the ground up" (173). He further elaborates the point thus:

> Start, in brief, from the standpoint of the subaltern; where "the subaltern" marks not a singular or essential subjectivity but a relational location from which to begin. Start with the concerns and experiences, categories and discourses, perceptions and problems of those groups visited by imperial and neoimperial imposition. Start from their perspectives, perceptions, and practices, and from there reconstruct social worlds. (173)

In short, through postcolonial relationism and perspectival realism, anticolonial approaches to sociological inquiry can yield important insights by forcing us to rethink already existing sociological concepts, categories, and theories from different perspectives. This will also necessarily lead sociologists to new concerns, interests, and questions. More importantly, however, it creates an opening for scholars to tear down the walls confining subjugated knowledge to the epistemic margins. As such, an anticolonial approach to environmental sociology has much to offer.

WHAT COULD AN ANTICOLONIAL APPROACH TO ENVIRONMENTAL SOCIOLOGY ENTAIL?

Of foremost importance, and at minimum, an anticolonial approach to environmental sociology hinges on the recognition that empire and colonial domination have not only shaped socioecological dynamics in the past but persist as elements that enable and influence socioecological dynamics in the present. We might consider this an ontological presupposition that makes such an approach to environmental-sociological inquiry possible in the first place. At the same time, as Go asserts, "It is not just about what happened in the world of empires but also about how empires have shaped how we see and understand the world; or what we do not see, what we do not understand" (2016:19). In other words, the anticolonial mode is furthermore presupposed by an epistemic commitment to knowing the world in ways that transcend and upend **coloniality**. Thus, an anticolonial environmental sociology should point to how and why our understanding of specific socioecological dynamics are limited by long-held (if unacknowledged) colonial assumptions, while also offering different ways of thinking about socioenvironmental phenomena. What might this entail?

First, it is important to note that an anticolonial approach is not reducible to simply writing about the environmental dimensions of imperial domination and colonialism. Environmental sociologists as divergent in thought as William Catton and Steven Bunker, among others, have written about colonialism in their respective analyses of society-environment interactions, yet neither of them could be understood as doing so with an anticolonial awareness.[2] In both instances, colonial domination is relegated to an important historical place and time, though no longer relevant role in the development of their respective socioecological objects of interest (i.e., the social and environmental consequences of extractive economies on the periphery of the capitalist world-system for Bunker and how Euro-Western society was able to overshoot environmental carrying capacity through colonial expansion for Catton). Moreover, in both instances, engagement with the topic of colonialism is subsumed by a broader theoretical framework that is firmly engrained within the **imperial/colonial episteme**. Although Bunker extends Marxist-inflected world-systems theory to account for the situation of the extreme periphery, he is nevertheless still committed to a historical materialist political economy, however modified. As for Catton, his discussion of colonial displacement is situated as merely a factor in the process by which modern society has come to overshoot the carrying capacity of the planet through population growth and the exuberant use of limited resources. In the end, neither engagement with colonialism expresses a commitment to challenging the ways that colonial rule has shaped how people think about and thus relate to each other and the more-than-human world.

There are at least two intellectual maneuvers that could inspire and/or facilitate anticolonial inquiry focused on the dynamics of socioecological phenomena. The first is to trace the ways that colonial processes influence or enable specific socioecological dynamics. The second is to theorize, or otherwise elucidate, socioecological phenomena in a way that emphasizes insights gathered from subaltern perspectives or standpoints. Of course, a particular project might focus on one or the other, or some combination of both, depending on the aims of the inquirer. Fortunately, there are many recent examples of research and writing by environmental sociologists that illustrate these two anticolonial gestures (see chapter 6 on settler colonialism). Note, however, that not all the examples cited below are explicitly anticolonial projects; nonetheless, they are illustrative.

Much of my own work in environmental sociology has drawn on both anticolonial gestures in projects focused on race. In "Ruin's Progeny: Race, Environment, and Appalachia's Coal Camp Blacks," my collaborators and I sought to disrupt the widespread narrative that links Blackness to environmental decay and suffering, which we felt characterized most of the research focused on Black communities in environmental sociology. Our phenomenological approach was based on the analysis of oral histories and other materials compiled by Eastern Kentucky African American Migration Project creator Karida L. Brown. The journal article demonstrates how:

> Despite living in a socionatural context fraught with problems, from social segregation to death from occupational hazards, [Black] people created meaning and value through engagement with myriad forms of non-human nature within the coal mining landscape. They found freedom in the same mountains that would later cause many community members an early death. In spite of the collapse of the coal economy and the subsequent ruination of the built landscape, Harlan and Letcher County's coal camp Blacks continue to return in celebration of their persistent connection to the land. (Brown et al. 2016, 340)

Such conclusions were made possible by starting with Black people's complex experiences and understandings of the natural world, rather than imposing a broader narrative onto their experiences of it. However, in this piece, we focused less on the colonial context in which the Black Appalachians found themselves.

By contrast, my current project is to illuminate the environmental significance of race using an anticolonial approach. This works points to the limits of prevailing modes of thinking about race and environment that I maintain are severely limited by colonial unknowing, or the onto-epistemological assumptions about race that deflect attention from the ongoing significance of colonial/imperial domination. By engaging with a range of social analysts often associated with Black studies and Indigenous studies, along with various other collaborators, I have sought to foreground the concerns, experiences, categories, and perceptions of the colonially subjugated to rethink race and environment from subaltern standpoints (Murphy 2021).

Other examples that perform either or both key intellectual maneuvers to varying extents abound. Hannah Holleman's work on the Dust Bowl provides a particularly salient example. Holleman (2017) reinterprets the Dust Bowl of the 1930s in a way that aligns with the vision of anticolonial environmental sociology by "provid[ing] an empirical and theoretical reinterpretation of what is frequently seen as an isolated historical-meteorological event, in order to address the wider social and ecological aspects of the crisis" (234). She critiques existing approaches to understanding the Dust Bowl that "make invisible the colonial and racial-domination aspects of the crisis [which] lead to the whitewashing of Dust Bowl narratives" by focusing solely on the ecological dimensions of the phenomenon and the ways in which they affected white settlers (235). Holleman's alternative approach reconnects colonial relations and moves beyond the imperial standpoint and its occlusions, thereby clarifying that:

> the disaster was one dramatic regional manifestation of a global socio-ecological crisis of soil erosion generated by the conditions of economic expansion via the "new imperialism" beginning in the 1870s and lasting through the early decades of the twentieth century[,] [which] include[d] policies and practices, such as the accelerated seizure of Indigenous lands, legitimated and spurred by a "culture of conquest" rooted in white supremacy. (234)

This example, along with many others, shows that environmental sociologists are already involved in the kind of work that could be considered as deploying core elements of an anticolonial approach.

To close, I want to further suggest that it is vital that scholars interested in an anticolonial approach to environmental sociology look beyond disciplinary boundaries, as has been a common practice by members of this epistemic community. Doing so is a necessary step toward any attempt to decolonize environmental sociology, especially considering that there are already spaces within the academy that have valued subaltern knowledge—

like Africana and Black studies, Native and Indigenous studies, colonial studies, as well as critical ethnic and racial studies, among other relevant transdisciplinary formations. Moreover, environmental sociologists like David Pellow, Robert Bullard, Lisa Park, and Dorceta Taylor have all held appointments in academic spaces like these at various turns throughout their careers, and I am currently a faculty member in a department of Black studies. The future for an anticolonial environmental sociology will depend on more transdisciplinary engagement, not less (see chapter 2).

NOTES

1. Catton (1980) uses the phrase "magnificent progress" to refer to the period of rapid growth in the American economy.
2. See Catton 1980, 26–29; Bunker 1984, 1023–29.

REFERENCES

Al-Hardan, Anaheed. 2022. "Empires, Colonialism, and the Global South in Sociology." *Contemporary Sociology* 51 (1): 16–27.

Bacon, Jules M. 2019. "Settler Colonialism as Eco-Social Structure and the Production of Colonial Ecological Violence." *Environmental Sociology* 5 (1): 59–69.

Bhambra, Gurminder K. 2014a. *Connected Sociologies*. Bloomsbury.

Bhambra, Gurminder K. 2014b. "Postcolonial and Decolonial Dialogues." *Postcolonial Studies* 17 (2): 115–21.

Bhambra, Gurminder K. 2020. "Colonial Global Economy: Towards a Theoretical Reorientation of Political Economy." *Review of International Political Economy* 28 (2): 307–22.

Bhambra, Gurminder K., and John Holmwood. 2021. *Colonialism and Modern Social Theory*. John Wiley & Sons.

Brown, Karida L., Michael W. Murphy, and Apollonya M. Porcelli. 2016. "Ruin's Progeny: Race, Environment, and Appalachia's Coal Camp Blacks." *Du Bois Review: Social Science Research on Race* 13 (2): 327–44.

Bunker, Steven G. 1984. "Modes of Extraction, Unequal Exchange, and the Progressive Underdevelopment of an Extreme Periphery: The Brazilian Amazon, 1600–1980." *American Journal of Sociology* 89 (5): 1017–64.

Burch, William R., Jr. 1971. *Daydreams and Nightmares: A Sociological Essay on the American Environment*. Harper & Row.

Cantzler, Julia Miller. 2020. *Environmental Justice as Decolonization: Political Contention, Innovation and Resistance Over Indigenous Fishing Rights in Australia, New Zealand, and the United States*. Routledge.

Carolan, Michael S. 2005. "Society, Biology, and Ecology: Bringing Nature Back into Sociology's Disciplinary Narrative Through Critical Realism." *Organization & Environment* 18 (4): 393–421.

Catton, William R., Jr. 1972a. Review of *Daydreams and Nightmares: A Sociological Essay on the American Environment*, by William R. Burch, Jr. *Social Forces* 51 (2): 240–41.

Catton, William R., Jr. 1972b. "Sociology in an Age of Fifth Wheels." *Social Forces* 50 (4): 436–47.

Catton, William R., Jr. 1980. *Overshoot: The Ecological Basis of Revolutionary Change*. University of Illinois Press.

Catton, William R., Jr. 2008. "A Retrospective View of My Development as an Environmental Sociologist." *Organization & Environment* 21 (4): 471–77.

Catton, William R., Jr., and Riley E. Dunlap. 1978. "Environmental Sociology: A New Paradigm." *American Sociologist* 13 (1): 41–49.

Connell, Raewyn. 2007. *Southern Theory: The Global Dynamics of Knowledge in Social Science*. Routledge.

Dunlap, Riley E. 2008. "Promoting a Paradigm Change: Reflections on Early Contributions to Environmental Sociology." *Organization & Environment* 21 (4): 478–87.

Dunlap, Riley E., and William R. Catton, Jr. 1979. "Environmental Sociology." *Annual Review of Sociology* 5:243–73.

Foster, Benjamin R. 2015. *The Age of Agade: Inventing Empire in Ancient Mesopotamia*. Routledge.

Freudenburg, William R., Scott Frickel, and Robert Gramling. 1995. "Beyond the Nature/society Divide: Learning to Think About a Mountain." *Sociological Forum* 10 (3): 361–92.

Freudenburg, William R. 2008. "Thirty Years of Scholarship and Science on Environment—Society Relationships." *Organization & Environment* 21 (4): 449–59.

Go, Julian. 2013. "For a Postcolonial Sociology." *Theory and Society* 42 (1): 25–55.

Go, Julian. 2011. *Patterns of Empire: The British and American Empires, 1688 to the Present*. Cambridge University Press.

Go, Julian. 2016. *Postcolonial Thought and Social Theory*. Oxford University Press.

Holleman, Hannah. 2017. "De-Naturalizing Ecological Disaster: Colonialism, Racism and the Global Dust Bowl of the 1930s." *Journal of Peasant Studies* 44 (1): 234–60.

Holleman, Hannah. 2018. Dust Bowls of Empire. Yale University Press.

Kumar, Krishan. 2020. *Empires: A Historical and Political Sociology*. Polity.

Manjapra, Kris. 2020. *Colonialism in Global Perspective*. Cambridge University Press.

McGee, Julius, and Patrick Trent Greiner. 2020. "Racial Justice Is Climate Justice: Racial Capitalism and the Fossil Economy." Hampton Institute, May 6.

Meghji, Ali. 2021. *Decolonizing Sociology: A Guide to Theory and Practice*. Polity.

Murphy, Michael Warren, and Caitlin Schroering. 2020. "Refiguring the Plantationocene: Racial Capitalism, World-Systems Analysis, and Global

Socioecological Transformation." *Journal of World-Systems Research* 26 (2): 400–415.

Murphy, Michael Warren, George Weddington, and AJ Rio-Glick. 2021. "Black Ecology and Critical Environmental Justice." *Environmental Justice* 14 (6): 393–97.

Murphy, Michael Warren. 2021. "Notes Toward an Anticolonial Environmental Sociology of Race." *Environmental Sociology* 7 (2): 122–33.

Murphy, Michael Warren. 2021. "On the Ecomateriality of Racial-Colonial Domination in Rhode Island." *Political Power and Social Theory* 38:161–89.

Norgaard, Kari Marie. 2019. *Salmon and Acorns Feed Our People: Colonialism, Nature, and Social Action*. Rutgers University Press.

Rice, James. 2013. "Further Beyond the Durkheimian Problematic: Environmental Sociology and the Co-Construction of the Social and the Natural." *Sociological Forum* 28 (2): 236–60.

Steinmetz, George. 2014. "The Sociology of Empires, Colonies, and Postcolonialism." *Annual Review of Sociology* 40 (1): 77–103.

Taylor, Dorceta E. 2014. *Toxic Communities: Environmental Racism, Industrial Pollution, and Residential Mobility*. NYU Press.

Taylor, Dorceta E. 2016. *The Rise of the American Conservation Movement: Power, Privilege, and Environmental Protection*. Duke University Press.

White, Damian, Alan Rudy, and Brian Gareau. 2016. *Environments, Natures and Social Theory: Towards a Critical Hybridity*. Palgrave Macmillan.

8. Ecofeminisms

Christine Labuski and Shannon Elizabeth Bell

"Climate Change is a man-made problem with a feminist solution." This sentiment, spoken by Mary Robinson, former president of Ireland and UN high commissioner for human rights, is the title of Robinson's 2019 book and the tagline of her popular three-season podcast, *Mothers of Invention*. Robinson has long argued that because women the world over are disproportionately vulnerable to climate change, it is imperative that they be centered in climate action efforts. Robinson's podcast, which she cohosts with Irish comedian and writer Maeve Higgens, primarily showcases the work of women who are working to mitigate the effects of climate change. But does a "feminist solution" to the climate crisis mean only a focus on women?

In this chapter, we argue that feminism—and ecofeminism in particular—can provide approaches to radical environmental thinking that move beyond a focus on the gender binary. We examine how ecofeminism's robust analyses of power and human supremacy are particularly valuable for understanding the intertwined ecological and social crises of our time, while also helping us imagine environmental futures that are racially and economically just, that promote multispecies flourishing, that value disabled, queer, and Indigenous lives, and that acknowledge the importance of socially reproductive—and feminized—forms of labor.

The concept of ecofeminism(s) is a decades-old social movement and theory that has much to say about the connections between gender, the environment, and climate change. Like the hosts of *Mothers of Invention*, we believe that climate and environmental justice strategies should be shaped by feminist thought; unlike the podcast, however, and as ecofeminists, we do not limit our understanding of "feminist" to any specific gender category or kind of person. Rather, and in the following discussion of ecofeminism's conflicted history, intersectional present, and possible futures, we understand **feminism** as *a practice of investigating and disrupting hierarchical relations of power.* Importantly, feminist thought and praxis are not limited to analyses of gender and/or sexuality. Rather, feminists develop their critiques of power via ongoing analyses of inequitable gender relations. In response, ecofeminist thinking strives to enact modes of accountability, humility, transparency, and collaboration that are antithetical to the inequitable distribution of power.

HISTORY IN BRIEF

Ecofeminism began, as both a theory and a movement, in the mid-1970s, and its central tenet is that there are noteworthy parallels between violent misogyny (and, by extension, homo- and transphobias) and humans' violent destruction of the earth. Specifically, ecofeminists see these violences as being rooted in, and enacted through, relations of domination, extractivism, and supremacy. In short, early ecofeminist thinkers argued that it wasn't possible to conceptualize an ecological movement without attending to these crosscutting modes of violent hierarchy. This broad analogy made it possible to link ecofeminist ideas and activism to a wide variety of international struggles, including food security, Indigenous environmental movements, animal rights, militarism, reproductive justice, gendered health disparities, and spirituality (Gaard 2011; Estévez-Saá and Lorenzo-Modia 2018).

The last of these connections became a source of vulnerability for ecofeminism's first wave, since it fed a perception that the movement was rooted in forms of "goddess worship" that were associated with *essentialist* thinking. In brief, **essentialism** is a way of linking all members of a certain group by imposing a set of natural traits onto them, a kind of "essence." For some critics, early ecofeminists' invocation of Mother Earth—and "her" need for protection—was a regressive move that "forced all women into the same 'feminine' category . . . of nurturers and caregivers" (Gonzales 2020), reinforcing a problematic and colonial set of gendered social norms from which many other feminists were eager to disassociate themselves.

The invocation of a spiritual and essentially feminized Mother Earth was also, at the time, strongly associated with wealthy white women who were detached from the on-the-ground environmental struggles of women of color. Indeed, noted environmental justice (EJ) scholar Dorceta Taylor (1997) argued that (white) ecofeminists paid more attention to animals than to racial inequalities and that their understanding of nature ignored the toxic urban environments endangering the health of communities of color: "Whereas ecofeminists recognize the degradation of nature and link it with the . . . devaluation of women," she argued, "women and men of color point to the degradation of nature, *then* to the accelerated . . . exploitation of nature in their communities . . . [arguing] that such degradation has a *racial and class basis*, too" (65; emphases added). Although this assessment was an accurate critique of some practitioners of ecofeminism, it did not reflect the totality of the movement (Gonzales 2020), as many early ecofeminists made explicit connections between people in "subordinate positions and the domination of nonhuman nature" (Warren 1994: 1), and many strove to link the simultaneous and entangled "oppressions of gender, ecology, race, species, and nation" (Gaard 2011, 128).

The introduction of intersectional feminist thought added texture to these conversations. Emerging in the United States in the early 1980s (Combahee River Collective 1981; hooks 1981; Crenshaw 1991), the theory of **intersectionality** centered the concerns of women of color in analyses of inequality by arguing that modes of oppression are concomitant and interlocking. "We . . . find it difficult," argued the Combahee River Collective (1981), "to separate race from class from sex oppression because in our lives they are most often experienced simultaneously" (213). Many environmental justice movements' ensuing commitments to intersectional praxis facilitated the incorporation of gender justice aims into their efforts (alongside their existing critiques of capitalism, colonialism, and racism), making the contributions of ecofeminism seem both narrow and redundant at the time (see Thomas, chapter 9, this volume).

Since the late 1990s, however, the proliferation of Black feminist scholarship—in combination with queer and trans theory and praxis—has sharpened and elaborated the patriarchal, binary-dependent, and heteronormative nature of environmental extraction and domination, underscoring the relevance of, and strongly influencing, contemporary ecofeminist thinking. From the sexual violence of fossil-fuel boomtowns (Labuski 2023), borne disproportionately by Indigenous women, to population-reduction proposals that ignore the grossly uneven contributions that families in global North nations have made to the climate crisis (Sasser 2016; Sultana 2022), to the **de-naturalizing** of queer and trans people via legislative violence, the issues around which contemporary ecofeminists are organizing reflect a refusal of social relations that jeopardize the well-being of marginalized communities and the more-than-human world.

Contemporary ecofeminism, then, reflects the scholarship and activism of the past several decades. As such, its **feminism** is defined primarily by critiques of widespread power relations, with gender functioning as just one entry point into these analyses (albeit one about which feminists have accumulated tremendous expertise). As did its precursors, contemporary ecofeminism attends to global North-South inequities, racism, ableism, homo- and transphobia, settler colonialism, and economic inequality. An important difference, however, is that contemporary ecofeminism interrogates the processes and institutions through which these structural inequalities are connected, most particularly through global capitalism's unrelenting drive to extract and accumulate surplus value from Nature and marginalized groups (Salleh 2017). As ecofeminist Ariel Salleh (2020) argues, the unrecognized and unpaid labor of what she terms "meta-industrial workers"—caregivers, peasant farmers, and Indigenous gatherers—"sustains, and indeed subsidises, the thermodynamic basis of capitalist patriarchal economies" (247).

Moreover, contemporary ecofeminists understand and insist that their movement "be led and articulated by frontline, impacted communities" (Feminist Agenda for a Green New Deal n.d.), i.e., those who are most threatened and/or harmed by ongoing forms of violent ecological domination. Importantly, these movements include reparations and restitution in their demands, with some scholars articulating this demand via feminist principles of accountability, collaboration, and humility (Sultana 2022; Gumbs 2020).

CURRENT CONVERSATIONS: DEGROWTH

Contemporary ecofeminism is also in conversation with other climate justice movements, including that of **degrowth**. Rooted in the twin concerns of resource depletion and widening economic inequality, the degrowth movement seeks to redirect society away from the "fetishism of growth" (Demaria and Latouche 2019) and instead construct a different kind of society, one in which there is "prosperity without growth" (Jackson 2009). The degrowth movement calls for the displacement of economic growth (and its proxy, the **Gross Domestic Product**, or **GDP**) as the primary indicator of national well-being. Given the demonstrable correlation between economic growth and the increased consumption of energy and material goods, the degrowth movement calls for an overall "scaling down and pulling back" (Crist 2019, 165) as a way to stabilize planetary temperature increases and to more equitably distribute the consequences of climate change. An explicit alternative to unrestricted capitalism, degrowth seeks to reconfigure economic wealth and to redirect resources toward communities and projects that are actively and demonstrably attempting to meet international climate justice goals (Hickel 2021). Crucially, a degrowth transition must be pluralist and **diverse**: as Demaria and Latouche (2019) argue, "it is not possible to formulate 'turnkey' solutions for degrowth" (150). Transition approaches must instead be tailored to the needs and geographies of specific regions and communities.

Consistent with intersectional feminist principles of collaboration and accountability, the degrowth movement also actively seeks redress for the disparate levels of responsibility—across the globe—for our current climate crises. Degrowth proponents propose not only the reparative payment of a **climate debt** from North to South, but also the advancement of *non-economic forms of growth*—e.g., health care, public transportation, and sustainable infrastructure—in regions and economic sectors where disparities are most acute. An ethos of **care and repair** has emerged alongside this movement, with (eco)feminists in particular enumerating how care can and should be redefined and deployed in a just transition. An economy aligned with feminist principles would have a "higher valuation for so-called pink labor jobs: caring, service, and regenerative labor, including everything from child and elder care to teaching, community art and events, and environmental rehabilitation, which could include caring for brownfields and other polluted ecosystems, especially those where vulnerable populations live" (Bell et al. 2020, 7). An ecofeminist vision for society would demand that care work is more highly valued, reimagined as visible, rewarded, and understood to be everyone's responsibility—not just that of women and racialized minorities.

Both the degrowth and care and repair movements challenge extractivist mentalities by explicitly questioning what constitutes *enough*. In this vein, proponents acknowledge that wealth and priority redistributions will lead to some forms of sacrifice, especially for those in possession of disproportionate material resources and power. However, as Shannon E. Bell, Cara Daggett, and Christine Labuski (2020) argue, these movements hold promise for many quality of life benefits as well. Although "wasteful and trivial kinds of commodity consumption will need to decrease, the focus should be upon what can be increased as a result: in slowing energy [and other forms of] consumption, we stand to gain in free time, justice, dignity, artisanship, community engagement, beauty, and rest" (Bell et al. 2020, 7). Moreover, a degrowth approach promises greater material equity and increased well-being for more people, by centering sufficiency and efficiency rather than profit and growth for growth's sake.

ECOFEMINISM'S INTELLECTUAL-POLITICAL TRAJECTORIES

Like other theoretical traditions, ecofeminism continues to evolve through being in conversation with other intellectual and political trajectories. As intersectional social movements, environmental and climate justice struggles center the perspectives of those communities that are structurally marginalized—often along multiple axes—and are most vulnerable to climate and other environmental threats. An intersectional approach to studying and/or acting within such communities, then, means attending to how these crises impact people "differently, unevenly, and disproportionately" (Sultana 2020, 118), and grappling with our "common but differentiated responsibility" (119) for effecting change. Far more than being an act of simple inclusion, acting on this awareness enlivens, energizes, and fundamentally transforms the goals and horizons for these movements. In this section, we outline a few of the issues most at stake for some of these intellectual and political communities, underscoring where and how ecofeminist thought intersects with their agendas.

Intersecting Communities: Indigeneity

Indigenous scholars and activists have long demonstrated the harms done by land dispossession and extraction to Indigenous peoples, communities, and to the more-than-human worlds with which these communities are in relation (Whyte 2018; Gilio-Whitaker 2020). A 2016 report coproduced by the Women's Earth Alliance and the Native Youth Sexual Health Network (WEA/NYSHN 2016) highlights how Indigenous women and young people, in particular, are victimized by forms of **environmental violence**—that is, "the disproportionate and often devastating impacts that the conscious and deliberate proliferation of environmental toxins and industrial development . . . have on Indigenous women, children and future generations, without regard from States or corporations for their severe and ongoing harm" (14). The report, titled "Violence on the Land, Violence on Our Bodies" (VLVB), draws clear and urgent connections between corporate greed, state collusion, settler colonialism, heteropatriarchal violence, and gender-based trauma, arguing that sexual violence against Indigenous women affects both land and cultural dispossession: "The reason women [are] attacked is because women carry our clans and . . . by carrying our clans, are the ones that hold that land for the next generation. That's where we get our identity as nations. So if you destroy the women, you destroy the nations, and then you get access to the land" (4).

Ecofeminists have much to learn from these concerns, including how to be in solidarity with these struggles while being respectful of the community-specific traumas and modes of healing to which they are connected. The VLVB report, for example, espouses an **anti-carceral** politics, seeking to avoid criminal punishment framings of environmental violence. Moreover, Indigenous environmental struggles framed by anti-settler perspectives and goals are at times anti-assimilationist, refusing to accommodate white and Western techno-optimisms seeking to further extract and profit from Indigenous knowledges. Arguing that "violence on the land *is* violence on our bodies," many strands of Indigenous environmentalism reflect a focus on forms of land restitution and the restored capacity to know and engage with the world in relational, rather than resource-inflected, ways (Kimmerer 2015; Gilio-Whitaker 2019; Puniwai 2020; Bacon and Vinyeta, chapter 6, this volume).

Intersecting Communities: Black and Women of Color Activism

Some of the first environmental justice activists were Black and Latina women, responding to polluting industries in their neighborhoods on behalf of their children and other vulnerable community members (Unger 2012). Since then, others seeking to connect race with environmentalism have framed calls for Black veganism, for example, via liberation and abolitionist sensibilities (Jones 2020), while others have outlined modes of solidarity between people marginalized by race and the increasingly endangered more-than-human world. In her book *Undrowned,* Black queer feminist Alexis Pauline Gumbs (2021) gleans a set of lessons from her deep engagement with marine mammals. Via poetically rendered themes of captivity, discovery, and the violent domination of human supremacy, Gumbs invites readers to consider the stakes of environmental degradation from the margins. In so doing, she echoes Black EJ thinker Mary Anäiise Heglar's (2019) argument that, for people of color, "climate change isn't the first existential threat." Crucially, both Gumbs and Heglar—in addition to the Indigenous authors of the VLVB report (above)—infuse their critiques with expressions of hope and repair, drawn from the frontline communities with which they are in conversation. White environmentalists hoping to "teach" communities of color about climate change, argues Heglar (2019), must keep in mind that such communities "have even more to teach *you* about building movements, about courage, [and] about survival" (emphasis added).

Black feminist ecological thought (Frazier 2020) is widely interdisciplinary and is attentive to the myriad ways that race and gender act as proxies for power relations across a host of environmental and climate issues. What is the relationship between the structural erasure of enslaved Black women's food and farming practices and the (very low) likelihood that a farm in the United States today will be owned by a Black woman (Williams 2021)? What role did the physical environment—of, for example, swamps and other Southern landscapes—play in how sexual violence and fugitivity intersected for self-emancipating Black women (Hyman 2021)? How do Black women experience fights for water in the United States, given this country's histories of segregated swimming pools, racist responses to disasters like Hurricane Katrina, and contaminated drinking water in cities with large Black populations, such as Flint, Michigan and Jackson, Mississippi (Henery 2021)? Black feminist scholars working in disciplines ranging from English literature to geography to public health are asking these and other vital questions, which many ecofeminist thinkers are learning from and engaging with.

Intersecting Communities: Queer and Trans Ecologies

Although essentialist ecofeminism "bore the marks" of conventional gender/sex binaries, using these binaries to "favor its own gender and ecological concerns" (Estévez-Saá and Lorenzo-Modia 2018, 126), such allegiances are out of place within ecofeminism's contemporary manifestations. Indeed, queer feminist and ecological thought is organized around the very real limitations of binary thinking, particularly that of *natural* and *unnatural*. Queer writer Alex Johnson (2011), for example, cites a twentieth-century compendium of "thousands" of instances of same-sex and sex/gender nonconformity among animals in order to suggest that any definition of "natural" that excludes queerness simply "betray[s] reality." Most recently, this binary has been weaponized by US politicians who have written and passed hundreds of pieces of legislation designed to render queer and trans people as wholly *unnatural*.

Early queer environmentalist thought focused on space and place, recognizing the important role that public parks and unfrequented urban areas played in cultivating queer connections, for example, through "cruising" (Mortimer-Sandilands and Erickson, 2010) and other forms of queer sociality made safer by these places' "out-of-the-way" locations. These histories unsettle idyllic conceptualizations of nature-spaces as *themselves* part of a binary pair, where parks are clean/white/nuclear-family-oriented and their opposite is an urban (read: racialized), contaminated, and dangerous Other. Some scholars, on the other hand, have used queer peoples' allegedly "unnatural" relationship to heteronormative reproduction to question the centrality of children and grandchildren to environmental discourse (Edelman 2004). As Noël Sturgeon (2009) has argued, relying on, rather than actively disrupting, this **repronormative** framework—that is, by organizing environmental activism solely around protecting future generations—risks supporting "the very relations of power that produce environmental problems" (12).

Some queer ecological thinkers argue that the sociopolitical hostility to which queer and trans people have long been subjected makes them especially potent climate justice activists. According to trans river scientist Cleo Wölfle Hazard (2022), "we queers and trans people know . . . transformations, change, and how to live fabulously while facing systemic attempts to destroy our life chances" (21). Relatedly, queer feminist science historian Michelle Murphy (2017) uses the term "**alterlife**" to describe not only the condition of bodies that have been permanently affected by environmental contaminants, such as PCBs, but also the promise and potential of such "altered" bodies to generate new futures via the new social relations that such "forever" chemicals make necessary. As Gumbs (2020) so poignantly asks, "What could enable us to live more porously, more mindful of the infinite changeability of our context, more open to each other and to our needs?" (61).

Intersecting Communities: Disability Justice Movements

Alongside their queer allies, disability justice activists demonstrate how disabled bodies and lives are rendered *unnatural*, typically by techno-optimistic narratives seeking to "fix" them (Shew 2023). They further point to the often hostile dimensions of their relationships to the natural world, from their extreme vulnerability in disaster scenarios (Engelman, Craig, and Iles 2022) to the numerous ways that their bodies are excluded from green spaces like parks and forests (Piepzna-Samarasinha 2021; Wrong 2021). Disability theory has also been mobilized by some scholars in order to think differently about remediation and repair; the concept of **disabled ecologies**, for example, allows us to consider the environment not solely as a set of problems to solve but also as an entity "requiring assistance, accommodation, and creative forms of care" (Taylor 2021).

As mentioned earlier, ecofeminists understand that such care work is vital for achieving environmental and climate justice. As the Feminist Agenda for a Green New Deal argues: "Our society is constructed upon and dependent on care work, [which] is valuable, low-carbon, community-based work that should be revalued." Disability studies highlights the *interdependent* nature of care work, viewing the need for and practice of care not as (feminized) weakness but rather as a collaborative space that effectively redistributes power relations and that can generate more habitable worlds. In contrast to the sometimes magical thinking of geoengineering, ecofeminists, in conversation with disability studies, start from an "understanding that nature has been altered and damaged in profound and serious ways" (Taylor 2021) and work to "stay with th[at] trouble" (Haraway 2016), even as they imagine how to heal it.

Intersecting Communities: Voices from the Global South

It is not uncommon to hear the refrain that women in the Global South are affected *first and worst* by climate crises. Ecofeminists who specialize in these regions argue that, in comparison to men, women "tend to rely more on 'nature' . . . and environmental resources . . . because of their weaker material rights and disproportionate care responsibilities" (Vercillo, Huggins, and Cochrane 2022, 1046). These realities often translate into women "having a greater interest in" the natural world (1046), as well as the anthropogenic harms to which it is subjected. It is crucial to note, however, the social dimensions of these gendered divisions of labor and how they can be compounded by the migration patterns of men seeking labor and other economic opportunities elsewhere. Ecofeminists, therefore, have embraced the "first and worst" refrain in order to draw attention to the very real gender imbalances with respect to how climate injustices are lived, but they are also careful to note that these vulnerabilities are structural, as they align with global economic arrangements rather than any inherent or *essential* gender/sex capacities. As Ariel Salleh (2010) argues, "Climate change, biodiversity loss, and social precarity are each results of capitalist overproduction" (205), and the "causes, effects, and solutions" to these crises are "heavily influenced by sex-gender norms" (Salleh 2020, 9).

An even more contentious issue among some eco-oriented feminists is that of **reproductive justice**. Catalyzed by a call from Global North feminist Donna Haraway (2016) to "make kin not babies" in order to limit population growth, feminists from the Global South (and their allies) have called attention to the limited accountability that Global South populations should be expected to assume, given their undersized contributions to climate crises. For many, such calls resonate acutely with the violent and racist histories through which their reproductive lives have been coercively managed. Women and families already burdened by climate injustices should not be asked to curtail their reproductive dreams by assuming an inequitable level of "embodied environmental responsibility" (Sasser 2016, 58).

But "making kin" can also be understood as a strategy for "welcoming limitations" (Crist 2019, 185) by humans who have overrun the earth's carrying capacity. The phrase also aligns with the goals of many queer activists, in that it prioritizes family and community relations that do not rely on heteronormative modes of reproduction, and it furthermore resonates with some Indigenous worldviews and practices through which members of the more-than-human world are welcomed as relatives rather than resources (Kimmerer 2015; Puniwai 2020). Indeed, Wölfle Hazard (2022) suggests that queer modes of kin-making are an avenue through which non-Indigenous humans can establish more intimate and loving relationships with a wider range of species. In short, feminists do not disagree about the histories of reproductive injustice that contour the concerns of Global South feminists, but they do not view population limitation discourses through the same frameworks; nor do they propose the same solutions.

Although each of these populations and justice movements bring distinct modes of awareness and solutions to the topics of environmental and climate justice, they all inspire ecofeminist movements and thinking. Contemporary ecofeminism strives to consider these modes together, alongside and entangled with one another, in a "multi-issue" way that can be messy, uncomfortable, and can even sometimes lead to conflict. We suggest, however, that this is the way forward, in that it can generate more kinds of solutions for more kinds of people and communities, including those that have yet to emerge. This is multi-issue and coalitional feminism at work.

CONCLUSION: ECOFEMINISM'S HORIZONS

As with other modes of feminist thought, it is best to think about ecofeminism in the plural—as multiple ecofeminisms, rather than as one unifying set of ideas. From this perspective, we can better appreciate the variety of voices—queer, disabled, Black, decolonial—whose climate and environmental justice concerns intersect with feminist critiques of patriarchy. Moreover, an ecofeminisms approach enables us to recognize the even wider set of ideas, practices, resistance movements, creative endeavors, and policy proposals that may not label themselves as (eco)feminist but with which we see ecofeminism in explicit conversation. As feminists committed to a plurality of environmental knowledges, we want to highlight these spaces of encounter and collaboration. Indeed, we are positively awestruck by the numerous projects, thinkers, activists, artists and more whose work we see in conversation with ecofeminism; we therefore close this chapter with just a few examples.

In her research with shark populations, Indigenous biologist Noelani Puniwai (2020) deploys what she calls a "methodology of *'aloha'* " (16) —

that is, a way of "relearning our relationships to our ocean relatives" (16). Puniwai defines "aloha" as "love," and in our reading of and engagement with ecofeminist theory and practice, we find **love** absolutely everywhere. Eco-critics Estevez-Saá and Lorenzo Modia (2018) agree, suggesting that love functions as a "key concept in ecofeminist writing" (143). Importantly, they continue, ecofeminist love is "evoked [not simply] as an abstraction, but rather as a practice arising from the ethical and aesthetical envisionings of eco-caring" (143).

For more than a decade, life partners Annie Sprinkle and Beth Stephens have used performance art as a mode of environmental activism, "marrying" the ocean, the sun and even a mountain in West Virginia. Sprinkle and Stephens call themselves **ecosexuals**, and use a mix of political resistance and playful creativity to call for greater intimacy with the natural world. By sensualizing—and even sexualizing—practices, such as walking barefoot, (naked) swimming, or having what they call "skygasms" (i.e., finding pleasure in gazing at clouds and the sun), they remind us of all that humans have to gain by dissolving the boundaries between human and nonhuman. This form of being "in love with" the natural and more-than-human worlds is poignantly expressed by queer ecological poet James McDermott (2022). For example, in his poem "My Queer Mind Goes for a Walk," he celebrates the freedom he feels when walking among animals in the woods who see him "as just another animal" (151).

Gumbs (2020) argues for the importance of what she calls "identification," which is the "process through which we expand our empathy" so that "the boundaries of who we are become more fluid because we identify with the experience of someone different" (8). That *someone*, however, doesn't have to be human. As Margarita Estévez-Saá and María Jesús Lorenzo-Modia (2018) argue, "continuit[ies] and connections" can and "should prevail between humans *and* non-human nature" (129; our emphasis). Moreover, the non-speciesism of ecosexuality resembles queer and Indigenous modes of kinship, which do not allow for sacrifice zones or disposable beings. Of paramount importance in all these relations is an understanding of how partial our knowledges(s) always are, and of the humility that comes with that understanding. Acknowledging the partiality—the incompleteness—of what we know underscores and opens us to our need for others. As Gumbs (2020) asks, "What if we are all necessary in our specificity?" (64).

In an interview published on the *Psychology Today* blog (Bekoff 2021), Gumbs points out that "[her] book probably says 'I love you' more than any other phrase," reminding her readers to think about the more-than-human world in these terms. In Lesson 19 of *Undrowned*, "Take Care of

Your Blessings," Gumbs's discussion of walrus tusks begins with the exhortation, "Love ice. Not war" (155). Indeed, Gumbs insists that even "hard" scientists love their objects of study, a sentiment with which Wölfle Hazard (2022) agrees: "Can we study extinction and contamination without shutting down our feelings of grief and loss?" They suggest that instead of trying to become numb to these feelings, we might rather "tap . . . into emotions as a source of social connection and political action" (153).

Love is just one emotion that we can center and deploy in our struggles for justice. "The coming decades of environmental crises will stretch not only toward death or health," argues Sunaura Taylor (2021), "but also [toward] something else—something impaired, precarious, dependent, filled with loss and struggle." Grief, anger, fear, passion—all these emotions have a place in the work of protecting, preserving, and fighting for the future of all life on earth. Rather than dismissing these emotions as "biases," ecofeminists recognize them as driving forces behind what we have chosen to study—and the motivations for maintaining a commitment to liberatory research and praxis.

REFERENCES

Beckoff, Marc. 2021. "*Undrowned: Black Feminist Lessons from Marine Mammals*: Alexis Pauline Gumbs tells us about her new book." *Psychology Today* (blog). www.psychologytoday.com/us/blog/animal-emotions/202101/undrowned-black-feminist-lessons-marine-mammals.

Bell, Shannon, Cara Daggett, and Christine Labuski. 2020. "Toward Feminist Energy Systems: Why Adding Women and Solar Panels is Not Enough." *Energy Research & Social Science* 68.

Combahee River Collective. 1981. "A Black Feminist Statement." In *This Bridge Called My Back: Writings by Radical Women of Color*, edited by C. Moraga and G. Anzaldúa. Persephone Press.

Crenshaw, Kimberlé. 1991. "Mapping the Margins: Intersectionality, Identity Politics, and Violence against Women of Color." *Stanford Law Review* 43 (6): 1241–99.

Crist, Eileen. 2019. *Abundant Earth: Toward an Ecological Civilization*. University of Chicago Press.

Demaria, Federico, and Serge Latouche. 2019. "Degrowth." In *Pluriverse: A Post-Development Dictionary*, edited by A. Kothari, A. Salleh, A. Escobar, F. Demaria, and A. Acosta. Tulika Books.

Edelman, Lee. 2004. *No Future: Queer Theory and the Death Drive*. Duke University Press.

Engelman, Alina, Leyla Craig, and Alastair Iles. 2022. "Global Disability Justice in Climate Disasters." *Health Affairs* 41 (10): 1496–1504.

Estévez-Saá, Margarita, and María Jesús Lorenzo-Modia. 2018. "The Ethics and Aesthetics of Eco-Caring: Contemporary Debates on Ecofeminism(s)." *Women's Studies* 47 (2): 123–46.

Feminist Agenda for a Green New Deal. https://feministgreennewdeal.com/.

Frazier, Chelsea Mikael. 2020. "Black Feminist Ecological Thought: A Manifesto." Atmos. October 1. https://atmos.earth/black-feminist-ecological-thought-essay/.

Gaard, Greta. 2011. "Ecofeminism Revisited: Rejecting Essentialism and Re-Placing Species in a Material Feminist Environmentalism." *Feminist Formations* 23 (2): 26–53.

Gilio-Whitaker, Dina. 2020. *As Long as Grass Grows: The Indigenous Fight for Environmental Justice, from Colonization to Standing Rock.* Beacon Press.

Gonzales, Laura Minton. 2020. "What is Ecofeminism? The Connection between Women and the Environment." Public Goods (blog). August 26. https://blog.publicgoods.com/what-is-ecofeminism/

Gumbs, Alexis Pauline. 2020. *Undrowned: Black Feminist Lessons from Marine Mammals.* AK Press.

Haraway, Donna. 2016. *Staying with the Trouble: Making Kin in the Chthulucene.* Duke University Press.

Harper, A. Breeze, ed. 2020. *Sistah Vegan: Black Women Speak on Food, Identity, Health, and Society.* Lantern.

Heglar, Mary Annäise. 2019. "Climate Change Isn't the First Existential Threat: People of Color Know All about Building Movements, Courage, and Survival." Zora. Medium. February 18. https://zora.medium.com/sorry-yall-but-climate-change-ain-t-the-first-existential-threat-b3c999267aa0.

Henery, Celeste. 2021. Presentation in "Black Feminist Ecologies." The Martyr's Shuffle. Wesleyan University. April 8, 2021. YouTube video, 2:18:32, here at 1:24:45. www.youtube.com/watch?v=lcIYr-mNwx8.

Hickel, Jason. 2021. *Less is More: How Degrowth Will Save the World.* Penguin.

hooks, bell. 1981. *Ain't I a Woman? Black Women and Feminism.* South End Press.

Hyman, Christy. 2021. Presentation in "Black Feminist Ecologies." The Martyr's Shuffle. Wesleyan University. April 8, 2021. YouTube video, 2:18:32, here at 1:15:08. www.youtube.com/watch?v=lcIYr-mNwx8.

Jackson, Tim. 2009. *Prosperity Without Growth: Economics for a Finite Planet.* Earthscan.

Johnson, Alex. 2011. "How to Queer Ecology: One Goose at a Time." *Orion Magazine,* March 24. https://orionmagazine.org/article/how-to-queer-ecology-once-goose-at-a-time/.

Kimmerer, Robyn Wall. 2015. *Braiding Sweetgrass: Indigenous Wisdom, Scientific Knowledge, and the Teachings of Plants.* Milkweed Editions.

Labuski, Christine. 2023. "Weeding Out the Weak: Labor, Gender, and Disability in a Fossil Fuel Boomtown." *Journal of Contemporary Ethnography* 52 (4): 559–83.

McDermott, James. 2022. "My Queer Mind Goes for a Walk." Special Issue, *Queer Ecologies, youarehere: the Journal of Creative Geography* 151.

Mortimer-Sandilands, Catriona and Bruce Erickson, eds. 2010. *Queer Ecologies: Sex, Nature, Politics, Desire*. Indiana University Press.

Mothers of Invention (podcast). https://www.mothersofinvention.online/.

Murphy, Michelle. 2017. "Alterlife and Decolonial Chemical Relations." *Cultural Anthropology*, 32 (4): 494–503.

Piepzna-Samarasinha, Leah. 2021. "Adaptive Joy." *Orion Magazine*, January 18. https://orionmagazine.org/article/adaptive-joy/.

Puniwai, Noelani. 2020. "Pua ka Wiliwili, Nanahu ka Manō: Understanding Sharks in Hawaiian Culture." *Human Biology* 92 (1): 11–17.

Salleh, Ariel. 2010. "From Metabolic Rift to 'Metabolic Value': Reflections on Environmental Sociology and the Alternative Globalization Movement." *Organization & Environment* 23 (2): 205–19.

Salleh, Ariel. 2017. *Ecofeminism as Politics: Nature, Marx and the Postmodern*. 2nd ed. Zed Books.

Salleh, Ariel. 2020. "A Materialist Ecofeminist Reading of the Green Economy: Or, Yes Karl, the Ecological Footprint Is Sex-Gendered." In *The Routledge Handbook of Transformative Global Studies*, edited by S.A. Hamed Hosseini, James Goodman, Sara Motta, and Barry K. Gills.. Routledge.

Sasser, Jade. 2016. "Population, Climate change, and the Embodiment of Environmental Crisis." In *Systemic Crises of Global Climate Change: Intersections of Race, Class and Gender*, edited by P. Godfrey and D. Torres. Routledge.

Shew, Ashley. 2023. *Against Techno-Ableism: Rethinking Who Needs Improvement*. W.W. Norton.

Sturgeon, Noël. 2009. *Environmentalism in Popular Culture: Gender, Race, Sexuality, and the Politics of the Natural*. University of Arizona Press.

Sultana, Farhana. 2022. "Critical Climate Justice." *Geographical Journal* 188 (1): 118–24.

Taylor, Dorceta E. 1997. "Women of Color, Environmental Justice, and Ecofeminism." In *Ecofeminism: Women, Culture, Nature*, edited by K. Warren. Indiana University Press.

Taylor, Sunaura. 2021. "Age of Disability: On Living Well with Impaired Landscapes." *Orion Magazine*, November 10. https://orionmagazine.org/article/age-of-disability/.

Unger, Nancy. 2012. *Beyond Nature's Housekeepers: American Women in Environmental History*. Oxford University Press.

Vercillo, Siera, Chris Huggins, and Logan Cochrane. 2022. "How Is Gender Investigated in African Climate Change Research? A Systematic Review of the Literature." *Ambio* 51:1045–62.

Warren, Karen J. 1994. "Ecological Feminist Philosophies: An Overview of the Issues." In *Ecological Feminist Philosophies*, edited by K. J. Warren. Routledge.

Whitworth, Lauran. 2019. "Goodbye Gauley Mountain, Hello Eco-Camp: Queer Environmentalism in the Anthropocene." *Feminist Theory* 20 (1): 73–92.

Whyte, Kyle Powys. 2018. "Settler Colonialism, Ecology, and Environmental Injustice." *Environment and Society* 9 (1): 125–44.

Williams, Teona. 2021. Presentation in "Black Feminist Ecologies." The Martyr's Shuffle. Wesleyan University. April 8, 2021. YouTube video, 2:18:32, here at 32:20.

Women's Earth Alliance/Native Youth Sexual Health Network. 2016. *Violence on the Land, Violence on Our Bodies: Building an Indigenous Response to Environmental Violence.* http://landbodydefense.org/.

Wölfle Hazard, Cleo. 2022. *Underflows: Queer Trans Ecologies and River Justice*. University of Washington Press.

Wrong, Yomi Suchiko. 2021. "At the Gate." *Orion Magazine,* January 18, 2022. https://orionmagazine.org/article/at-the-gate/.

9. Intersections of Environmental Justice

Tanesha A. Thomas

The environmental justice (EJ) framework recognizes that environmental hazards disproportionately impact marginalized communities that have intersecting identities and social forces contributing to their vulnerability. EJ advocates for all people to have safe, healthy environments and the right to meaningfully participate in decisions that affect their well-being. Environmental injustices affect human beings unequally along the lines of race, gender, class, and nation, so an emphasis on any one of these will dilute the explanatory power of any analytical approach (Pellow and Brulle 2005). "It is not enough, then, for any analysis to focus on a limited number of components of environmental injustice without acknowledgment of wider structures of power and prejudice" (Buckingham and Kulcur 2009, 71). Therefore, scholars and policymakers cannot understand or prevent environmental injustices through a singularly focused framework that emphasizes one form of inequality to the exclusion of others. An intersectional analysis is key to truly understanding and dismantling environmental injustice; it demands a more integrated consideration of issues, rather than a single-axis or single-issue-based analysis.

Critical legal scholar Kimberlé Crenshaw's work on the concept of intersectionality has had a particularly salient impact on analyzing multiple forms of oppression in society (1989, 1990). Crenshaw's work on intersectionality demonstrated that women of color were at a distinct disadvantage within the American legal system, which treated issues of gender and race discreetly, rather than as overlapping and reinforcing concerns. Existing at the intersection of race and gender, women of color failed to achieve justice; their cases fell through the cracks of the legal system built to acknowledge gendered and racial discrimination, but not both.

Building on Crenshaw's work, scholars have applied intersectionality more broadly to examine "the relationships among multiple dimensions and modalities of social relations and subject formations" (McCall 2005, 1771). *Intersectionality* acknowledges that all parts of human identity, including gender, race, sexuality, ability, heritage, ethnicity, socioeconomic status, spirituality, legal status and much more, are inseparable and interconnected. "These inequalities are mutually reinforcing in that they tend to act together to produce and maintain systems of individual and collective power, privilege, and subordination" (Pellow 2018, 19).

EJ scholarship is increasingly embracing intersectionality in what EJ scholars David Pellow and Robert Brulle (2005) call critical EJ. A critical EJ framework captures the intersectional nuances of environmental burdens, moving society closer to true social and EJ. A second generation of EJ scholarship is using intersectionality to advance "environmental justice research as a tool for more accurately identifying and challenging contextual intersections of power in order to achieve environmental justice and equity" (Malin and Ryder 2018, 3). This emerging field of research contributes to a more interdisciplinary, more diverse, and more justice-oriented environmental sociology. This theoretical framework also advances society toward a more sustainable future for us all.

ENVIRONMENTAL RACISM

Environmental racism is a particular type of environmental injustice. It can take the form of racial discrimination in environmental policymaking and enforcement of regulations and laws, the deliberate targeting of communities of color for toxic waste facilities, the official sanctioning of the presence of life-threatening poisons and pollutants in communities of color, or the history of excluding people of color from leadership within the environmental movement (Bullard 1996, preface). In 1982, events in Warren County, North Carolina forced the topic of environmental racism onto the national agenda. Residents and supporters of this mostly Black community joined in protesting the construction of a burial site for soil contaminated with highly toxic polychlorinated biphenyls (PCBs). Protestors lay down in front of dump trucks, literally putting their bodies on the line to stop pollution from entering their community. Hundreds were arrested in the first instance of people being jailed for trying to halt a toxic waste landfill in US history (Bullard 1990; Dowie 1995).

Although the trucks still rolled in, this event inspired a movement of research about environmental racism. The United Church of Christ's (UCC)

Commission on Racial Justice published the landmark 1987 study, *Toxic Wastes and Race in the United States*. The UCC found that race was the most significant factor in determining where commercial hazardous waste treatment, storage, and disposal facilities (TSDFs) were located in the United States (Commission 1987). In 1992, the US General Accounting Office (GAO) and the Environmental Protection Agency's (EPA) conducted an EJ analysis of the Southern United States. Although the report did not use the terms "EJ" or "environmental racism," the results clearly indicated such; throughout the South commercial hazardous waste facilities were disproportionately located in majority Black neighborhoods, which also contained high numbers of Black residents living below the poverty level (GAO 1992).

Since these inaugural studies, EJ scholarship has discovered ways that race intersects with other forms of identity to produce environmental injustices for people of color. For example, scholars have examined how houselessness disproportionately impacts people of color. Houselessness is an environmental justice issue. African Americans make up more than 40 percent of the US houseless population and Native American people more than 4 percent, while only accounting for about 13.6 percent and 1 percent of the total population, respectively (Olivet et al. 2018). Without proper shelter, people without housing are directly exposed to natural weather conditions. This can increase the likelihood of adverse health outcomes that would otherwise be avoidable if such people had more secure housing and/or economic conditions. These dangers are compounded by the threat of criminalization. Law enforcement patrolling spaces where houseless people gather often force them into toxic spaces or dangerous living conditions with insects, rodents, or other hazards that can threaten human health (Goodling 2020). Examining the intersecting threats and hazards houseless people of color face can help policymakers better formulate laws to ensure just and affordable housing for all.

EJ scholars are also increasingly articulating the ways that race intersects with capitalism to oppress people of color (Pulido 2015, 2017). The Flint water crisis is one of the most highly publicized examples of how poverty intersects with environmental racism (Robinson et al. 2018). Flint, Michigan is primarily a poor Black community. In 2015, the median household income there was $24,679, and the city had a shocking poverty rate of 41.6 percent, 167 percent higher than the national average. Black residents constitute a "surplus population," lacking secure employment, making them expendable to the capitalist economy (Robinson 2013). Michigan governor Rick Snyder and the municipal emergency fiscal manager (EFM) changed the city's water source from the Detroit River to the Flint River to save money in 2014. Black people in Flint constituted only 14 percent of

Michigan's population, yet they were approximately 49 percent of those living under the EFM's policy. Poor Black residents were disproportionately vulnerable to decisions made by the EFM creating the conditions for environmental racism (Pulido 2016).

It is common knowledge that the Flint River was contaminated by emissions from the General Motors factory. As the polluted water traveled through the city's crumbling plumbing infrastructure, toxins leached into the city's water supply, impacting the taste, smell, and appearance of the water. Environmental health testing revealed alarming levels of lead in the water: seven to ten thousand times over the federal legal limit. Despite these well-known hazards, the state moved forward with the plans to use the Flint River as the city's water supply. The health of a primarily Black, low-income community was not a priority for the state. Under the conditions of racial capitalism, Flint residents were rendered disposable owing to the intersection of their racial and economic status (Pulido 2016). Understanding how structures of oppression such as racism and capitalism intersect and reinforce oppression can help understand and prevent future environmental tragedies.

Race can intersect with an individual's legal status or place in the occupational structure to also produce environmental racism. Farm work is an environmental racism issue. The US farmworker population is overwhelmingly Latino, speaks little English, and at least half are undocumented (meaning the workers lack legal immigration status or permission to work in the United States). Employers exploit the fact that Latino workers are at risk of deportation and therefore less likely to report occupational abuses, specifically pesticide exposure (Berkey 2017; Lincoln 2018). Farmworkers are exposed to unknown quantities of pesticides, which are widely used in US agriculture. Agricultural pesticides contain chemicals to protect produce from insects, certain plants, fungi, and rodents. However, these toxins can be devastating for human health, causing a variety of health issues from rashes, headaches, nausea, and vomiting, to respiratory failure, coma, cancer, neurologic or reproductive problems, and even death. Farmworkers and their families are exposed to pesticides at work, at home, and at school because the toxins travel through the air beyond the areas targeted for application as dust or droplets during and after application (Arcury et al. 2002). Figure 9.1 is a photo of Alfredo "Lelo" Juárez at the 2024 Marcha Campesina protest in Mt Vernon, Washington. Juárez is a member of the group Community 2 Community, which is based in Bellingham, Washington. Community 2 Community is an eco-feminist social justice organization led by women of color. In Figure 9.1, people are holding signs reading "Fair wage for spring

FIGURE 9.1. Farmworker organizer Alfredo "Lelo" Juarez and others at the 2024 Marcha Campesina in Mt. Vernon, Washington. Photo credit: Sattva Photo, courtesy of Community to Community, Bellingham, WA.

flowers!" and "40 hour guaranteed! Saldos justos!" The protestors and their messaging reflect the overlapping interests of environmental justice with other issues such as economic security. Groups like Community 2 Community highlight how people's lived experiences necessitate an intersectional approach to solving social problems.[1]

COLONIAL ECOLOGICAL VIOLENCE

The intersection of race and indigeneity makes Native, Indigenous, and First Nations peoples uniquely vulnerable to environmental injustice (see chapters 6 and 7). The term "**Indigenous peoples**" refer to the descendants of the original inhabitants in a particular geographic area. The colonization and environmental destruction of Indigenous peoples have gone hand in hand. A Native American woman from North Dakota was quoted as saying, "Ever since the white man came here, they keep [pushing us back, taking our lands, pushing us onto reservations. We are down to three percent now, and I see this as just another way for them to take our lands, to completely annihilate our races. We see that as racism" (Krauss 1994, 267). This statement describes how environmental racism intersects with capitalism and patriarchy to produce colonial ecological violence. Colonial ecological violence is the result of settler-colonialism and environmental racism that oppress Native and Indigenous peoples around the world (Bacon 2019).

During her acceptance speech at the 2015 Goldman Environmental Prize ceremony Honduran environmental activist Berta Cáceres stated: "Let us wake up, humankind! We're out of time. We must shake our conscience free of the rapacious capitalism, racism and patriarchy that will only assure our own self-destruction." Berta helped organize a grassroots movement with the Indigenous Lenca communities across western Honduras to challenge the construction of a hydroelectric dam on the Gualcarque River. Tragically, Berta Cáceres was murdered in her home by armed intruders in 2016. She had reason to suspect the hit men were hired by Desa, the dam construction company (Lakhani 2020). Cáceres was bringing attention to "an intersectionality that destroys bodies and worlds, the intersectionality of structures of subordination that killed her" (Méndez 2018). Unfortunately, her death is part of a long history of colonial ecological violence.

Berta Cáceres died trying to protect the Gualcarque River, which is sacred to the Lenca. Most of the water powering dams, factories, industrial parks, mines, and other capitalist industrial projects comes from the ancestral rivers and wells of Indigenous communities (Mackey 2016; Gedicks 2001). These water sources are seen as exploitable sources of energy in the eyes of the capitalist-development complex; however, they have significantly different meanings for Indigenous peoples, whose livelihoods and worldviews are intimately tied to them. These massive industrial projects "result in the violent displacement of peoples from their sources of material well-being, but also seek to cut off the lifeblood that nurtures Indigenous cosmovisions" (Méndez 2018). These forces of colonial ecological violence threaten the livelihoods and survival of Indigenous peoples around the world.

Various social institutions and scholars are including frameworks based on intersectionality and Indigenous philosophies, ontologies, and epistemologies such as relationality (McGregor et al. 2020). Relationality recognizes "that we, the land, the water, and all living creatures are related and, as relatives, we are meant to love and care for each other" (Wilson 2015, 255). The Hopi and the Navajo, two Tribes that have traditionally occupied the Southwestern United States, have beliefs that are diametrically opposed to the Western capitalist social system that dominates them from the outside; these groups believe that the earth cannot be owned. They view themselves as caretakers of the earth, called to live a simple life in harmony with nature. Maintaining harmony through daily practices and prayers that reinforce these linkages are key to Indigenous spirituality. The Western capitalist ethic and value system is quite different. Many Anglo-Americans view the land and its resources as a commodity to be bought and sold for profit. This philosophy seeks to transform ecosystems rather than living in harmony with them (Hall 1994).

Researchers are incorporating the Indigenous concept of relationality to improve environmental and sustainability education (ESE). One of the field's biggest challenges is the exclusion of contributions from women and non-Western peoples. In response, ESE is deepening its intersectional analysis by incorporating ecofeminism, queer pedagogy, biocentric ethics, and Indigenous and decolonizing perspectives (Maina-Okori et al. 2018). ESE is just one area where investigating how environmental justice intersects with indigeneity can help us bridge the gap between Indigenous and Western knowledge systems (Levac et al. 2018).

An intersectional EJ is necessary to address the exceptional environmental challenges facing Native Peoples around the world (Olsen 2018; Jamieson et al. 2023). Too often, Indigenous peoples are treated as a monolith, in the process ignoring internal power structures and dynamics that may exist within each group. Intersectional analysis furthers our understanding of the nuanced experiences of women and other marginalized intra group members (Osborne et al. 2019), particularly, the rampant and epidemic of violence against Indigenous women and girls. Women from these communities suffer from physical, emotional, and sexual violence at epidemic rates. Gender, race, and indigeneity makes these women's oppression less visible to mainstream society. These communities struggle for recognition and resources to address the specific challenges they face as poor Native women (Maddison and Partridge 2014; Clark 2016; García-Del Moral 2018; Horowitz 2017; Guimont et al. 2020). In another example, researchers in Canada found that Indigenous youth possess an intersectional resilience despite various hardships (e.g., abandonment, victimization, racism), difficulties, and/or periods of sustained environmental stress (e.g., poverty, discrimination). These youth still maintained a strong determination and will to pursue education in the face of distress, a motivation to be responsible, to lead exemplary lives, and to influence others positively. Researchers found that the youth demonstrated three dominant intersecting themes: (1) strengthening cultural identity and family connections; (2) engagement in social groups and service to self and community; and (3) practices of the arts and a positive outlook (Njeze et al. 2020).

ECOFEMINISM

The events in Love Canal, New York put a female face to the environmental justice movement. In 1978, children in this blue-collar housing development started complaining to their mothers that their feet burned when they played barefoot outside. Residents soon found out that Love Canal had been knowingly built on a former toxic waste site. State epidemiologists tested the population and found elevated levels of headaches, birth defects, miscarriages, epilepsy, liver disease, and rectal bleeding. Although the state declared that the situation constituted a serious threat to health and safety of local residents, it was slow to follow up and address the community's health concerns. When EPA agents arrived to inspect the area in 1980, housewife Lois Gibbs and other residents held them "hostage" inside one of their homes for hours demanding a commitment to action. Two days later, US President Jimmy Carter agreed to purchase residents' homes in the contaminated area and relocate the residents to safer spaces. Women like Gibbs are often called to action by threats to their families and communities. Love Canal, "an all-American, white, middle-class suburb-with its poisoned children and irate mothers-struck a chord in the American imagination" (Dowie 1995, 129).

Mothers of all races and ethnicities, including the Mothers of East Los Angeles (MELA), were fighting for the health and safety of their families. This Mexican American EJ group gained prominence in the 1980s for successfully organizing against the construction of a prison and an incinerator. The group was underestimated because they were poor Latinas. MELA successfully lobbied state legislators, who were "very surprised that a group of mothers from East Los Angeles knew so much of what was happening, of the laws. They were expecting that once these persons would be asked a question, they would not know how to answer. Every question they were asked, the Mothers of East Los Angeles had answers for" (Gutierrez 1994, 225).

MELA represented a growing but underacknowledged aspect of the mainstream environmental movement: the gendered factors that impact one's environmental privilege or environmental burdening (Buckingham and Kulcur 2009). The gender-based division of labor in society makes women primarily responsible for family health, particularly children. Women are often the first to make the connection between toxic hazards and human health when they see the effects in their family. For women, environmental activism is directly linked to good parenting. MELA member Aurora Castillo was quoted as saying, "before the first Anglo set foot on California soil, we were compelled to unite, because the future quality of life for our children is being threatened. You know if one of [her] children's safety is jeopardized, the mother turns into a lioness" (Gutierrez 1994, 223).

Without stating it explicitly, women like Lois Gibbs and MELA engage in their own kind of ecofeminism. Ecofeminism addresses the connections between the oppression of women and the exploitation of the environment; it also provides a strong analytical foundation to challenge patriarchal oppression of women and nature (Maina-Okori et al. 2018). Ecofeminism acknowledges how gender intersects with other forms of oppression, including racism, classism, ableism, heterosexism, and patriarchy, to create environmental injustices that uniquely impact women and girls.

Race intersects with gender to produce different starting points for and subsequent analysis of environmental injustice for women of color. Their traditional role of mothers transcends their public role in the community as protectors of the race. For example, Black mothers see their activism within EJ movements as challenges to traditional racial stereotypes of Black women as lazy and unintelligent (Krauss 1994, 265). Native American women do not separate their identities from their commitment to the environment: "Our connection as women is to the Mother Earth, from the time of our consciousness. We're not environmentalists. We're born into the struggle of protecting and preserving our communities. We don't separate ourselves. Our lifeblood automatically makes us responsible; we are born with it" (Krauss 1994, 269). Indigenous writer and activist Patricia Monture-Angus articulated it as follows: "To artificially separate my gender (or any other part of my being) from my race and culture forces me to deny the way I experience the world" (1996). Intersectionality allows us to understand the nuanced relationships that diverse types of women have with EJ.

In her analysis of EJ campaigns in New York City, Julie Sze (2006) found that gender and race can intersect to vilify mothers of color. Bronx County has among the highest incidences of asthma emergency department visits and hospital discharges both in the city and throughout the state (New York State 2009). Environmental officials neglected scientific data that pointed to industrial pollution as a cause for asthma in the Bronx, instead choosing to focus on poor housekeeping practices. Despite evidence of both indoor and outdoor causes of asthma, the government initiated a program targeting the housekeeping practices of low-income mothers of color. Outdoor causes of asthma and respiratory illness, including pollution from the expressway and idling diesel buses, were disregarded. The focus shifted away from inadequate state regulation to mothers' allegedly dangerous parenting.

Women of color can face unfair environmental burdening in the workplace as well. A study of nail salon workers in California revealed that women of color face "unnecessary occupational and everyday exposures to compounds such as the aromatic amines used in dyes, pesticides and plastics." Salon workers, such as nail technicians, experience intersecting identities as undocumented laborers, women, people of color, immigrants, and non-Native English speakers. Employers target and exploit them to produce an "ethnic caste system and poverty trap." As a result, poor Asian women are disproportionately exposed to toxic chemicals, resulting in respiratory and skin ailments that may lead to intergenerational effects via pregnancy (Jampel 2018). Like pesticides, these toxins may follow workers home and continue to harm family members and others around them for generations.

Environmental disasters and the recovery response to them are also shaped by the intersection of gender and race (Boyce 2000; Ryder 2017; Luft 2016; Weber and Messias 2012). For example, in the aftermath of Hurricane Katrina, young Black women found themselves caught at the intersection of race, gender, and age. As Black people they were more likely to live in houses that suffered the most damage from the storm. As women, they were more likely to bear the burden of caregiving children and family members as households broke up, causing them to endure stress and anxiety. As young people these new physical and mental challenges were more disabling owing to their inability to work, lacking younger children to help with household tasks, and more limited resources (Jampel 2018).

QUANTIFYING INTERSECTIONAL ENVIRONMENTAL JUSTICE

From this foundation in intersectional theory, scholars are forging new intellectual paths by applying quantitative methods, such as multivariate linear models, conventional multilevel analysis, and multilevel analysis of individual heterogeneity and discriminatory accuracy (MAIHDA) (Bauer et al. 2021; Gross and Goldan 2023). Below are some examples of researchers applying these and other quantitative methods to evaluate intersectional environmental health risks to communities.

Raoul Liévanos (2015) used spatial and demographic analysis to find that race, gender, social class, citizenship, and language all intersect to uniquely impact Latine communities. The study found that the intersection of high concentrations of low-income and foreign-born Latines, Latina single-mother families, and primarily Spanish-speaking households was the strongest predictor of exposure to carcinogenic air pollution in 2005. Liévanos (2019) updated the study to include Indigenous people, finding that the concentration of disadvantaged Indigenous peoples was a significant

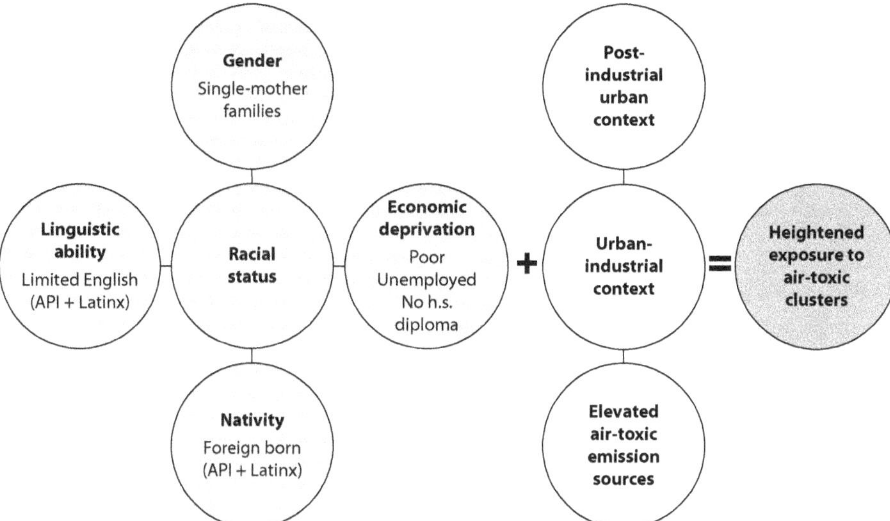

FIGURE 9.2. Predictors of exposure to carcinogenic air pollution. Source: Liévanos 2019, 163; reproduced with permission.

positive predictor of exposure to carcinogenic air pollution in the Mid-Atlantic region (see fig. 9.2).

Patricia Hill Collins and colleagues (2011) conducted a similar analysis of intersecting factors increasing cancer risks from air toxics along the US-Mexico border. They found that Latines' ethnic status intersects with class, gender, and age status to amplify a disproportionate cancer risk. Sara Elizabeth Grineski and colleagues (2013) looked at differential exposure among different Latine groups in Miami, Florida. They found that after disaggregating the Latine population based on country of origin, that Cubans and Colombians were more at risk for cancer compared to Mexicans who faced a significantly decreased risk.

Applying a unique quantitative method, they call eco-intersectional multilevel modeling, Camila H. Alvarez and Clare Rosenfeld Evans (2021, 2022) find that neighborhoods with higher concentrations of people of color, single-mother households, below-average income, and educational attainment in metropolitan areas experience drastically higher levels of exposure to carcinogens in the air. Combined, these factors create a unique exposure to cancer risk.

The environmental justice movement started focusing on how racism produced patterns of environmental harm in Black American communities.

Researchers have since found that environmental racism is exacerbated by capitalism and the criminal justice system. Applying Kimberlé Crenshaw's theory of intersectionality (1989, 1990), they have expanded the field of environmental justice to include the analysis of how various social identities can impact environmental vulnerability. This chapter covers only a few of the areas that environmental sociologists have explored, including race, indigeneity, gender, ethnicity, citizenship, and language. Applying an intersectional analysis to environmental justice has exposed unique environmental vulnerabilities for certain groups like undocumented farmworkers exposed to pesticides, Indigenous people protecting sacred land and water, mothers fighting to protect the health and safety of their families, and nail technicians exposed to toxins in the workplace. These are a few of the social identities that have been found to intersect and increase environmental vulnerability. The studies summarized in this chapter demonstrate that a more interdisciplinary, diverse, and justice-oriented environmental sociology ensures sustainability and social welfare for our collective future.

NOTE

1. Alfredo "Lelo" Juárez was arrested by Immigration and Customs Enforcement on March 27, 2025. In July of 2025, after spending almost four months in harsh conditions at the Northwest Detention Center in Tacoma, WA, he voluntarily left the United States, as the hurdles to fight an immigration case in the country were too steep. Community to Community has been campaigning for his release. The author of this chapter and the book editors would like to express solidarity with Alfredo Juárez and the work done by C2C to secure his release, as well as with all unfairly imprisoned and deported migrants and those supporting them.

REFERENCES

Alvarez, Camila H., and Clare Rosenfeld Evans. 2021. "Intersectional Environmental Justice and Population Health Inequalities: A Novel Approach." *Social Science & Medicine* 269:113559.

Alvarez, Camila H., Anna Calasanti, Clare Rosenfeld Evans, and Kerry Ard. 2022. "Intersectional Inequalities in Industrial Air Toxics Exposure in the United States." *Health & Place* 77:102886.

Arcury, Thomas A., Sara A. Quandt, and Gregory B. Russell. 2002. "Pesticide Safety among Farmworkers: Perceived Risk and Perceived Control as Factors Reflecting Environmental Justice." *Environmental Health Perspectives* 110 (2): 233–40.

Bacon, Jules M. 2019. "Settler Colonialism as Eco-Social Structure and the Production of Colonial Ecological Violence." *Environmental Sociology* 5 (1): 59–69.

Bauer, Greta R., Siobhan M. Churchill, Mayuri Mahendran, Chantel Walwyn, Daniel Lizotte, and Alma Angelica Villa-Rueda. 2021. "Intersectionality in Quantitative Research: A Systematic Review of Its Emergence and Applications of Theory and Methods." *SSM-Population Health* 14:100798.

Berkey, Rebecca E. 2017. *Environmental Justice and Farm Labor*. Routledge.

Boyce, James K. 2000. "Let Them Eat Risk? Wealth, Rights and Disaster Vulnerability." *Disasters* 24 (3): 254–61.

Buckingham, Susan, and Rakibe Kulcur. 2009. "Gendered Geographies of Environmental Injustice." *Antipode* 41 (4): 659–83.

Bullard, Robert D. 1990. *Dumping in Dixie*. Westview Press

Bullard, Robert D. 1994. *Unequal Protection: Environmental Justice and Communities of Color*. Sierra Club Books.

Bullard, Robert D. 1996. "Environmental Justice: It's More than Waste Facility Siting." *Social Science Quarterly* 77 (3): 493–99.

Clark, Natalie. 2016. "Red Intersectionality and Violence-Informed Witnessing Praxis with Indigenous Girls." *Girlhood Studies* 9 (2): 46–64.

Collins, Timothy W., Sara E. Grineski, Jayajit Chakraborty, and Yolanda J. McDonald. 2011. "Understanding Environmental Health Inequalities through Comparative Intracategorical Analysis: Racial/Ethnic Disparities in Cancer Risks from Air Toxics in El Paso County, Texas." *Health & Place* 17 (1): 335–44.

Commission for Racial Justice. 1987. *Toxic Wastes and Race in the United States: A National Report on the Racial and Socio-Economic Characteristics of Communities with Hazardous Waste Sites*. United Church of Christ.

Crenshaw, Kimberlé. 1989. "Demarginalizing the Intersection of Race and Sex: A Black Feminist Critique of Antidiscrimination Doctrine, Feminist Theory and Antiracist Politics." *University of Chicago Legal Forum* 1989 (1): 139–67.

Crenshaw, Kimberlé. 1990. "Mapping the Margins: Intersectionality, Identity Politics, and Violence against Women of Color." *Stanford Law Review* 43 (6): 1241–99.

Dowie, Mark. 1995. *Losing Ground: American Environmentalism at the Close of the Twentieth Century*. MIT Press.

GAO (US General Accounting Office). 1992. *Siting of Hazardous Waste Landfills and Their Correlation with Racial and Economic Status of Surrounding Communities*. Government Printing Office.

García-Del Moral, Paulina. 2018. "The Murders of Indigenous women in Canada as Feminicides: Toward a Decolonial Intersectional Reconceptualization of Femicide." *Signs: Journal of Women in Culture and Society* 43 (4): 929–54.

Gedicks, Al. 2001. *Resource Rebels: Native Challenges to Mining and Oil Corporations*. South End Press.

Goodling, Erin. 2020. "Intersecting Hazards, Intersectional Identities: A Baseline Critical Environmental Justice analysis of US Homelessness." *Environment and Planning E: Nature and Space* 3 (3): 833–56.

Grineski, Sara Elizabeth, Timothy W. Collins, and Jayajit Chakraborty. 2013. "Hispanic Heterogeneity and Environmental Injustice: Intra-Ethnic Patterns of Exposure to Cancer Risks from Traffic-Related Air Pollution in Miami." *Population and Environment* 35:26–44.

Guimont Marceau, Stéphane, Dolores Figueroa Romero, Vivian Jiménez Estrada, and Roberta Rice. 2020. "Approaching Violence against Indigenous Women in the Americas from Relational, Intersectional and Multiscalar Perspectives." *Canadian Journal of Latin American and Caribbean Studies* 45 (1): 5–25.

Gutierrez, Gabriel. 1994. "Mothers of East Los Angeles Strike Back." In Bullard, *Unequal Protection*.

Hall, Kathy. 1994. "Impacts of the Energy Industry on the Navajo and Hopi." In Bullard, *Unequal Protection*.

Horowitz, Leah S. 2017. " 'It Shocks Me, the Place of Women': Intersectionality and Mining Companies' Retrogradation of Indigenous Women in New Caledonia." *Gender, Place & Culture* 24 (10): 1419–40.

Jamieson, Lisa, Xiangqun Ju, Dandara Haag, Pedro Ribeiro, Gustavo Soares, and Joanne Hedges. 2023. "An Intersectionality Approach to Indigenous Oral Health Inequities: The Super-Additive Impacts of Racism and Negative Life Events." *PLoS One* 18 (1): e0279614.

Jampel, Catherine. 2018. "Intersections of Disability Justice, Racial Justice and Environmental Justice." *Environmental Sociology* 4 (1): 122–35.

Krauss, Celene. 1994. "Women of Color on the Front Line." In Bullard, *Unequal Protection*.

Lakhani, Nina. 2020. *Who Killed Berta Cáceres?: Dams, Death Squads, and an Indigenous Defender's Battle for the Planet*. Verso.

Levac, Leah, Lisa McMurtry, Deborah Stienstra, Gail Baikie, Cindy Hanson, and Devi Mucina. 2018. *Learning across Indigenous and Western Knowledge Systems and Intersectionality: Reconciling Social Science Research Approaches*. Unpublished SSHRC Knowledge Synthesis Report, University of Guelph.

Liévanos, Raoul S. 2015. "Race, Deprivation, and Immigrant Isolation: The Spatial Demography of Air-Toxic Clusters in the Continental United States." *Social Science Research* 54:50–67.

Liévanos, Raoul S. 2019. "Air-Toxic Clusters Revisited: Intersectional Environmental Inequalities and Indigenous Deprivation in the US Environmental Protection Agency regions." *Race and Social Problems* 11 (2): 161–84.

Lincoln, Elizabeth. 2018. "Accountability for Pesticide Poisoning of Undocumented Farmworkers." *Hastings Environmental Law Journal* 24 (2): 383–412.

Luft, Rachel E. 2016. "Racialized Disaster Patriarchy: An Intersectional Model for Understanding Disaster Ten Years after Hurricane Katrina." *Feminist Formations* 28 (2): 1–26.

Mackey, Danielle Marie. 2016. "An Interview with Gustavo Castro, Sole Witness to Assassination of Berta Cáceres." *The Intercept*, April 18.

Maddison, Sarah, and Emma Partridge. 2014. "Agonism and Intersectionality: Indigenous Women, Violence and Feminist Collective Identity." *Intersectionality and Social Change* 37:27–52.

Maina-Okori, Naomi Mumbi, Jada Renee Koushik, and Alexandria Wilson. 2018. "Reimagining Intersectionality in Environmental and Sustainability Education: A Critical Literature review." *Journal of Environmental Education* 49 (4): 286–96.

Malin, Stephanie A., and Stacia S. Ryder. 2018. "Developing Deeply Intersectional Environmental Justice Scholarship." *Environmental Sociology* 4 (1): 1–7.

McCall, Leslie. 2005. "The Complexity of Intersectionality." *Signs: Journal of Women in Culture and Society* 30 (3): 1771–1800.

McGregor, Deborah, Steven Whitaker, and Mahisha Sritharan. 2020. "Indigenous Environmental Justice and Sustainability." *Current Opinion in Environmental Sustainability* 43:35–40.

Méndez, María José. 2018. " 'The River Told Me': Rethinking Intersectionality from the World of Berta Cáceres." *Capitalism Nature Socialism* 29 (1): 7–24.

Mohai, Paul, Paula M. Lantz, Jeffrey Morenoff, James S. House, and Richard P. Mero. 2009. "Racial and Socioeconomic Disparities in Residential Proximity to Polluting Industrial Facilities: Evidence from the Americans' Changing Lives Study." *American Journal of Public Health* 99 (S3): S649–S656.

Monture-Angus, Patricia, and Suzanne M. Stiegelbauer. 1996. "Thunder in My Soul: A Mohawk Woman Speaks." *Resources for Feminist Research* 25 (1/2): 52.

New York State Department of Health. 2009. *New York State Asthma Surveillance Summary Report*. Public Health Information Group and Center for Community Health.

Njeze, Chinyere, Kelley Bird-Naytowhow, Tamara Pearl, and Andrew R. Hatala. 2020. "Intersectionality of Resilience: a Strengths-Based Case Study Approach with Indigenous Youth in an Urban Canadian Context." *Qualitative Health Research* 30 (13): 2001–18.

Olivet, Jeffrey, Marc Dones, Molly Richard, Catriona Wilkey, Svetlana Yampolskaya, and Maya Beit-Arie. 2018. *Supporting Partnerships for Anti-Racist Communities (SPARC): Initial Findings from Quantitative and Qualitative Research*. Center for Social Innovation.

Olsen, Torjer A. 2018. "This Word Is (Not?) Very Exciting: Considering Intersectionality in Indigenous Studies." *NORA-Nordic Journal of Feminist and Gender Research* 26 (3): 182–96.

Osborne, Natalie, Catherine Howlett, and Deanna Grant-Smith. 2019. "Intersectionality and Indigenous Peoples in Australia: Experiences with Engagement in Native Title and Mining." In *The Palgrave Handbook of Intersectionality in Public Policy*, edited by O. Hankivsky and J.S. Jordan-Zachery. Palgrave Macmillan.

Pellow, David N. 2018. *What Is Critical Environmental Justice?* Polity.

Pellow, David N., and Robert J. Brulle. 2005. *Power, Justice, and the Environment: A Critical Appraisal of the Environmental Justice Movement.* MIT Press.

Pulido, Laura. 2015. "Geographies of Race and Ethnicity 1: White Supremacy vs White Privilege in Environmental Racism Research." *Progress in Human Geography* 39 (6): 809–17.

Pulido, Laura. 2016. "Flint, Environmental Racism, and Racial Capitalism." *Capitalism Nature Socialism* 27 (3): 1–16.

Pulido, Laura. 2017. "Geographies of Race and Ethnicity II: Environmental Racism, Racial capitalism and state-sanctioned violence." *Progress in Human Geography* 41 (4): 524–33.

Robinson, Cedric J. 2013. "Race, Capitalism, and the Antidemocracy." In *Reading Rodney King/reading urban uprising,* edited by R. Gooding-Williams. Routledge.

Robinson, Tomeka M., Garrett Shum, and Sabrina Singh. 2018. "Politically Unhealthy: Flint's Fight against Poverty, Environmental Racism, and Dirty Water." *Journal of International Crisis and Risk Communication Research* 1 (2): 303–24.

Ryder, Stacia S. 2017. "A Bridge to Challenging Environmental Inequality: Intersectionality, Environmental Justice, and Disaster Vulnerability." *Social Thought & Research* 34:85–115.

Sze, Julie. 2006. *Noxious New York: The Racial Politics of Urban Health and Environmental Justice*. MIT press.

Weber, Lynn, and DeAnne K. Hilfinger Messias. 2012. "Mississippi Front-Line Recovery Work after Hurricane Katrina: An Analysis of the Intersections of Gender, Race, and Class in Advocacy, Power Relations, and Health." *Social Science & Medicine* 74 (11): 1833–41.

Wilson, Alexandria. 2015. "A Steadily Beating Heart: Resistance, Persistence and Resurgence." In *More Will Sing Their Way to Freedom: Indigenous Resistance and Resurgence,* edited by E. Coburn. Fernwood Press.

PART III

Transformations

10. The Corporation in Environmental Sociology

J. P. Sapinski

The way of life in "modern" countries is created, first and foremost, by corporations.

SOLÉ 2009, 82; my translation

INVISIBLE CORPORATIONS

In this chapter, I argue that environmental sociologists ought to consider corporations in their work. Why talk—again—about corporations? We know they're the "bad guys," that they're the ones screwing up the environment, right? True, a lot of attention is paid to corporations in the media and popular discourse. And activists certainly know how to target specific corporations to great effect. Yet, are corporations really a core element of environmental sociological analysis? A quick survey of publications in environmental sociology is informative. In the two main journals of the field, *Society and Natural Resources* and *Environmental Sociology*, a search for the keyword "corporation" in the abstracts of papers published between 2012 and 2024 returns, respectively, seven out of 1,286 articles for the former, and four out of 385 articles for the latter. The latest editions of some key textbooks of the discipline—Michael Meyerfeld Bell and colleagues' *An Invitation to Environmental Sociology* (2020), Michael Carolan's *Society and the Environment* (2020), and Kenneth A. Gould and Tammy L. Lewis's *Twenty Lessons in Environmental Sociology* (2021)—do acknowledge corporations as key actors of environmental destruction. Yet their coverage of the corporation *as a topic of environmental sociological inquiry* varies substantially, and only Gould and Lewis include a chapter dedicated to the topic, with a narrow focus on the corporate media (Campbell 2021).

This underemphasis on the corporation is not surprising, nor is it specific to the subfield. Bastiaan van Apeldoorn and Naná de Graaf (2017) make the same observation for the discipline of political science as a whole, where they find analysis of the role of the corporation is uneven at best. French sociologist

Andreu Solé (2011) goes further. He argues that the corporation is nothing less than a blind spot of all social sciences, especially when it comes to environmental debates. We do see the corporate presence everywhere. For most of us living in modern "Western" contexts, we are born in hospitals that in the United States often belong to corporations, and we spend our lives working for corporations (or trying to work for them, or supporting a partner who works for them) to earn the money necessary to buy from other corporations the **commodities** we depend on to eat, move around and communicate. When we die, we are most often buried or cremated by corporations (Abraham 2016). Even our interpersonal relations have been made into a commodity that social media corporations sell to advertisers. Yet that corporate presence is so all-encompassing that we stop paying attention to it; we don't question it anymore; we stop thinking about it and accept it as a fact of life. This is why Solé (2011) argues that the corporation is the dominant institution of today's world: it is a totalitarian institution whose ubiquitous existence means it becomes invisible, like the water we swim in, or even the fish tank itself; it is the unquestioned force that shapes our world and bounds our very existence (28).

In this chapter, I argue that environmental sociology needs to analyze the corporation as the key institution that it is. I assert that the subdiscipline needs to move beyond pointing the finger at corporations for destroying the ecosystems, the water, and the atmosphere we depend on, and to produce an actual sociological analysis of the institution behind this destruction, and this for three reasons. First and foremost, as I will show below, the corporation is *the* institution that mediates the exchanges between human societies and their environment—their **social metabolism**—under advanced capitalism. Second, viewing the corporation as the crucial social institution that it is provides a much better theoretical understanding of the socioecological processes that are the object of environmental sociology. Such focus sheds light on who the actors of environmental destruction are, how they operate, and why this destruction apparently sees no end, despite all the corporate social responsibility goals or environmental, social, and governance principles put in place in individual firms and other organizations. Third, shedding light on these processes directly supports socio-environmental movements working to stop destruction. Critical environmental sociologists have produced an excellent analysis of how capitalism as a mode of production implies environmental destruction. But, as sociologist Yves-Marie Abraham points out, what socio-environmental activists deal with in their daily struggles is not abstract capitalism, an entity too vague to be grappled with; it is individual corporations, and the corporation as a social institution (Abraham 2016).

In what follows, I sketch out a framework that puts the corporation where it belongs in environmental sociology: front and center. After an historical overview of how the corporation became a central institution of the capitalist mode of production, I detail how it mediates human social metabolism. I conclude with brief remarks on the discourse that corporations have constructed that legitimizes and even glorifies this mediation role, and by discussing how emphasizing the corporation might conversely bring attention to alternative institutions and discourses that foreground a different, non-corporate future.

A BRIEF HISTORY OF THE CORPORATION

Generally speaking, a "corporation" (from the Latin *corpus*, body) is simply a group of individuals legally recognized as a distinct entity, established for a stated purpose, which exists beyond its individual members. Groups of people have incorporated for many millennia for various purposes, as religious bodies, teaching institutions, municipalities, and so on, to provide institutional stability and independence from its members (Stern 2017). The **legal personhood** granted to a corporation allows it to enter into contracts, to defend its interests in courts of law, and to be protected from its members' liabilities, just as its members are also protected from the corporation's liabilities. In what follows, I discuss a specific type of corporation, the business corporation, established for the private profit of its members.

The Early Corporation

The business corporation has not always been the totalitarian institution it has become today. Historians trace early corporate forms to the temporary agreements among merchants to pool capital and split risks from long-distance trading voyages found in most ancient societies, including Assyria, China, and the Roman Empire (Davoudi et al. 2018; Stern 2017). The precursor to the modern corporate form, the state-chartered company, emerged as an instrument of colonization during the period of European expansion in the sixteenth and seventeenth centuries. Imperialist monarchies viewed it as a means to mobilize merchant and, in some cases, aristocratic capital on a long-term basis to develop new trade routes, fund new settlements, militarily conquer local populations, and fend off competing colonial states. In a continent still dominated by the church, the chartered company became the main form of international trade organization, already accounting for the majority of world trade in the sixteenth century (Taylor and Baskerville 1994). States gave companies monopolies over trade routes and over large tracts of unceded Indigenous land so that they could establish extractive industries, including the fur and fish trade based on Indigenous procurement in colder climates, and slave labor-based mining and sugar cane and cotton plantations in warm areas (Ferdinand 2021; Taylor and Baskerville 1994). Trade and extractivism benefited both colonial states, which were able to secure access to land and resources to consolidate their power in the face of other states, and companies' wealthy sponsors by returning a profit on their investments. Prefiguring modern-day corporations, some of these early corporations were structured as joint stock companies, governed by a corporate charter and a board of directors, and issuing tradeable shares to investors (Stern 2017). However, beyond industry and commerce, powers granted by the state additionally included taking possession of Indigenous territories, nominating colonial governors, and waging war against Native populations and other European states (Brandon 2015).

As far back as the sixteenth century, chartered companies were thus key institutions for the integration of "new" territories into the expanding Euro-centered world economy. They hence presided over dramatic changes in the political ecology of these lands. In what is now North America, the intensified commercial pressure on Indigenous trade, especially the fur and fish trade in the seventeenth and eighteenth centuries (Taylor and Baskerville 1994), led to lasting changes in the relationship to the land. With the increase in the volume of catch and furs destined to European markets, animal populations declined—quite dramatically in the case of the beaver. Further south, the slave trade led by Dutch and English chartered companies had deep socio-ecological impacts on African kingdoms and peoples, as well as on Indigenous groups, as Native landscapes were replaced by plantations (Ferdinand 2021). More importantly, Europeans brought to the Americas a different conception of the relationship to the land based on private property, which allowed and justified continued extraction of resources for trade, opposite to Indigenous views that emphasized kinship with the land and its creatures (Ferdinand 2021; Savard 1980), a worldview that today is more than ever embodied in how the corporation operates.

The Modern Corporation

The modern corporation emerged in the late nineteenth century. Since then, it has reorganized much further relationships among humans, and between humans and the land. Starting in the mid-nineteenth century in the United States and Britain, several characteristics derived from the earlier chartered company form were codified into law by state managers. This series of legal changes institutionalized the corporate form that had previously required state approval on a case-by-case basis. Capital requirements were rapidly increasing as industry moved to fossil fuels as a prime mover, as railways expanded to transport ever greater quantities of raw materials and manufactured goods, and as steel-making grew in scale to meet concomitant demand. Limited liability—limiting the amount of corporate debt each shareholder is responsible for to the amount invested in the venture—that states previously granted separately was coupled to incorporation, first in England in 1856 and then progressively in US states. By making it safer to invest in stockholding, this move mobilized middle-class savings to fund industrial projects at an increased scale (Bakan 2004). Likewise, as the number of incorporation requests grew rapidly, state managers decided to streamline the process of legislatively chartering each individual incorporation into a bureaucratic process to alleviate pressure on the legislative process. Such innovations provided business corporations with a strong legal status, and started the process of institutionalizing the corporation as the main form commercial entities would later take (Prechel 2000; Stern 2017). What's more, the economic turmoil of the second half of the nineteenth century led capitalists to seek consolidation in most industry branches. At the time, legislative action was required in each case to allow corporate ownership of other corporations (Prechel 2000). The modern form of the corporation was created between 1888 and 1899, when, in the face of pressures to attract capital to increase revenue, legislators of the state of New Jersey passed a series of amendments to its incorporation laws. These amendments allowed all corporations incorporated in the state to own stock in other corporations, including those incorporated out of state. Hence, according to economic sociologist Harland Prechel (2000), at the end of the nineteenth century, New Jersey legislation had "established the legal basis for the modern corporation" (34). Given structural pressures to attract capital, similar laws were rapidly adopted elsewhere in the United States, the Anglo-Saxon world, and the world over.

One last principle that underlies the modern corporation was institutionalized with the 1919 *Dodge v. Ford* Michigan Supreme Court judgment (Bakan 2004). Based on precedents going back to the last decades of the previous century (Davoudi et al. 2018), this judgment essentially codified into law the "best interests of the corporation" principle, according to which a corporation and its directors' foremost duty is to maximize shareholder wealth (Bakan 2004). Since then, legal scholars such as Joel Bakan view the corporation as legally mandated to return profit, and thus to **externalize** costs as much as possible.

The socio-environmental consequences of creating the modern corporation, and thereby allowing industrial consolidation on a scale unseen before, cannot be understated. As a legal person, a corporation does not have a physical existence, and its interests are often at odds with those of so-called "natural persons"—regular human beings—or nonhuman life. Yet it is granted various rights, including the right to free speech to further its political interests, and the right to sue citizens or states who threaten its activities or its profit-making ability. Widespread limited liability protects shareholders from being held responsible for the corporation's debt, including that incurring from the social and environmental damages it causes. Unlimited stock ownership in other corporations allows a single entity to control vast swathes of capital with a relatively small investment, thus greatly increasing the scale of capital accumulation (Carroll and Sapinski 2018; Prechel 2000) and thereby the scale of material and energy throughput. This concentration of business operations and thus of revenue led to a vastly increased political power. Just a few centuries ago, state-chartered companies were a tool of state power; now, governments have become beholden to large corporations for tax revenue and job provision to their constituents. Powers that had been exceptional privileges throughout history became in the late nineteenth century attributes of all corporations, making the corporation *the most powerful institution of our times.*

In sum, the historical period of European colonization that saw the emergence of the capitalist mode of production witnessed at the same time the institutionalization of the business corporation, from a restricted arrangement to facilitate trade, and later, colonial expansion, to the dominant institution of today's world. Sociologically, the corporation emerged from complex relationships between micro-, meso-, and macro-dynamics involving businesspeople and state managers at the bottom, and the structures of capitalism and interstate competition at the top (fig. 10.1). In the middle, the corporation acts as the agent of capital—a collective capitalist—at the same time as it enables the actions of some individuals and constrains those of others. Analyzing the corporation thus introduces the meso-level of complexity within discussions of structure and agency fundamental to social thought since at least C. Wight Mills (1959). All corporations, business and others, exert agency as meso-level groups, formed in support of micro-level individual pursuits within macro-level dynamics. State-chartered companies were created and given agency by individual decisions made within a structural context of imperial rivalries, where they gave a strategic advantage to those monarchs bent on increasing their power. After the capitalist mode of production had replaced mercantile expansionism, another institutional form was gradually created, again by individual lawmakers' decisions in response to state-level problems. These decisions that

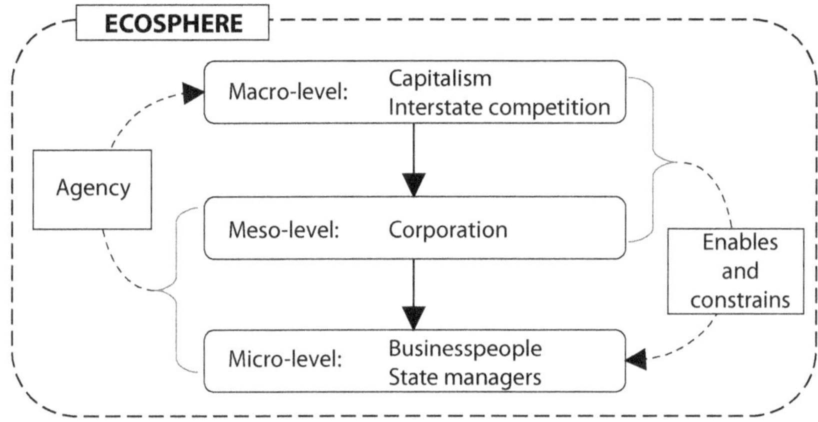

FIGURE 10.1. Structure, agency, and the corporation. Source: author, from Ken Hatt (personal communication, 2011).

increased the corporation's breadth of action *set in motion new dynamics where corporate interests gained autonomy and came to prevail over all others,* thereby determining at the structural level future directions of capitalism. It is obvious to all that environmental destruction is one consequence of the now autonomous corporation playing out its interests within these complex multi-level dynamics. In the next section, I will further argue that it is the corporation that now mediates the metabolic relationship between humans and the land, to its own benefit.

CORPORATE METABOLISM

Since its inception in the late nineteenth and early twentieth centuries, the modern corporation mediates social metabolism—the fundamental exchanges of matter and energy between human societies and their environment (Pineault 2023)—in capitalist society. Thinking in terms of commodity chains helps understand this mediation. The expression "**commodity chain**" is used to talk about all the stages involved in producing a commodity. It was first coined by Terence K. Hopkins and Immanuel Wallerstein (1977) to emphasize the cross-border linkages involved in the production of all commodities in the current world economy, as well as the actors involved at each stage. These stages go from the extraction of raw materials from the earth's crust or the biosphere in different sites across the world, their multiple transformations in factories near and far, and their final assemblage in yet other locations into end consumer products (Bair

2009). Commodity chains also bring the attention to the multiple impacts of capitalist production on the land as raw materials are successively transformed into final commodities: each stage of a chain takes place somewhere, and hence involves relationships with the ecosphere and impacts on the land (Bridge 2008).

Take one of today's mundane commodities, a smartphone. The chain begins when mining corporations such as the Swiss Glencore or Arizona-based Freeport-McMoRan, and oil giants like Shell or ExxonMobil are granted claims on vast tracts of land by the state. These claims legally allow them to commodify parts of the earth's crust as they extract and refine the various metals and the petroleum products that enter into the production of electronic devices. Processed metals and plastics are then transferred to other corporations that build the various components of smartphones. In Chinese factories, Taiwan's Foxconn and other manufacturers receive these components and direct their workers to assemble the units following specifications received from Apple, Samsung, and so on, which plan the production process and market end commodities directly to consumers or through various retailers such as Amazon or Walmart. Large financial corporations, banks, and investment companies, underwrite operations across the chain, and all these actors purchase insurance from yet another set of corporations. Finally, at the very end of the chain, waste management corporations collect and transport the devices that are deemed not to be useful anymore to where they are disassembled and decommodified.

From the perspective of this chapter, commodity chains first draw attention to the specific institution in charge of each stage of these chains in various locations of the globe: the corporation. Whether operating globally, like Apple or ExxonMobil or large banks and insurance companies, or as national and regional subcontractors, like Foxconn, corporations are the main, if not the only, actors across global commodity chains. Second, looking at the commodity chain and its corporate actors brings to light the impacts of all stages of production on the land. Analysts and activists generally emphasize the beginning and end points of the chain—extraction and disposal—as the sites of contact between the production process and the land. Ecological destruction by the oil and mining extractive industries has been targeted since a long time (e.g., Carroll 2021; Deneault and Sacher 2012), as has been waste disposal (e.g., Liboiron and Lepawsky 2022; Sicotte 2016). Yet, in the middle of the chain, the Foxconn factories, beyond requiring large tracts of land to sit on, are built using steel and concrete whose production has considerable environmental consequences, requires energy to operate, and generates sizable waste streams in the atmosphere and in landfills; oil refineries are major polluters and among the largest greenhouse gas emitters.

Third, and most importantly for my argument, as corporations are the actors of each stage of production, their agency and the structural constraints they are embedded in determine how the relationship to the land is conducted. In a capitalist mode of production, production units are engaged in competition against each other. They must reduce the cost of each unit produced so they can either sell cheaper than their competitors and capture their market share or return a higher profit and reinvest in further efficiency and/or growth to decrease costs even more. More profitable corporations will also attract more investors and benefit from lower interest rates, enhancing their capacity to mobilize capital for more efficient production and growth (Carroll and Sapinski 2018; Schnaiberg 1980). Competition therefore drives growth, even in economic sectors dominated by a small number of large corporations. In this context, there is great pressure to "externalize" social and environmental costs—to ensure these costs are paid not by the corporation but by public money or by ordinary citizens (Bakan 2004). As an example of **externalities**, electronic waste—discarded smartphones and other electronic devices—is often transported to sites located in the Global South, where informal waste pickers use fire or acid to separate the precious metals from other components and sell them for a living. The process creates noxious fumes that workers and residents of nearby areas inhale, directly threatening their health (Lebbie et al. 2021). These people and the health system they rely on are the ones paying the costs of treating this waste, instead of the manufacturers who then get to reduce costs and increase profits.

The structural context also pushes corporations to spend large amounts to lobby government officials and state managers for more permissive regulation or to avoid passing regulation that would reduce corporate profit (Carroll and Sapinski 2018). The well-documented campaign to sow doubt on the climate crisis is a case in point (Oreskes and Conway 2010). Legal provisions for limited liability effectively shield the corporate managers and directors who decide on corporate strategy from being held accountable for the costs and impacts borne by other people, other organizations, and the land—including its nonhuman inhabitants. As to the concept of corporate personhood, it has been used, at least in the United States, to remove any limitation on corporate electoral spending, considered as "free speech" protected under the First Amendment since the *Citizens United v. Federal Electoral Commission* 2010 Supreme Court decision. Hence, US corporations can spend as much as they want to elect those who support their

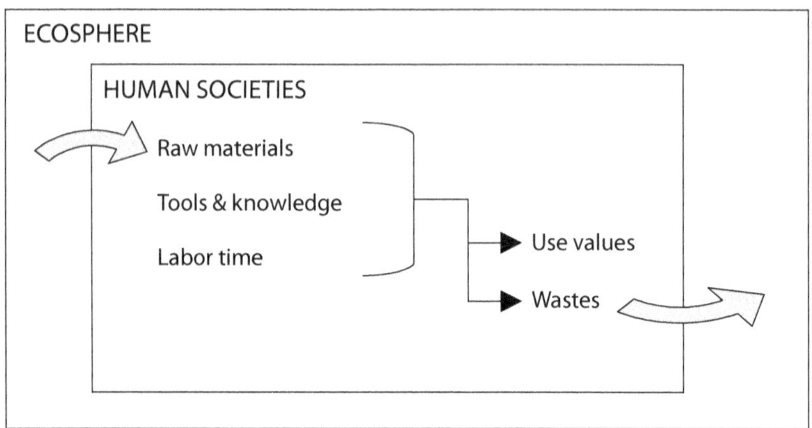

FIGURE 10.2. General form of the social metabolism of human societies. Source: author.

capacity to return the highest profits and externalize the most costs. Overall, the corporate form created in the late nineteenth and early twentieth centuries thus allows business interest in search of higher returns to determine how the land will be used. The land, for the corporation, is a site of unlimited extraction of various living and nonliving matters destined to be turned into commodities eventually sold to consumers, with the sole purpose of returning the highest profit for corporate shareholders, and with no consideration for alternative uses by other human and nonhuman beings.

All societies, in a similar way to individual organisms, exist in a metabolic relationship with the land (Pineault 2023). They draw from it the living and nonliving materials they need to maintain their existence, transform them, and distribute them to their constituent individual members. The general form of this relationship under noncapitalist modes of production is summarized in figure 10.2: humans spend labor time to extract raw materials and to use their knowledge and their tools to transform them into use values—the objects they use in their everyday life—from food and shelter to artwork and ceremonial apparatus, returning to the land as waste the unusable leftovers from the process. Under the capitalist mode of production, since its institutionalization, *the corporation inserts itself as the social institution within which the social metabolism takes place* (fig. 10.3).

It is the corporation that mobilizes human labor time to extract the raw materials it decides to extract and that become the corporation's property, and to transform them using the tools and machines it owns, according to

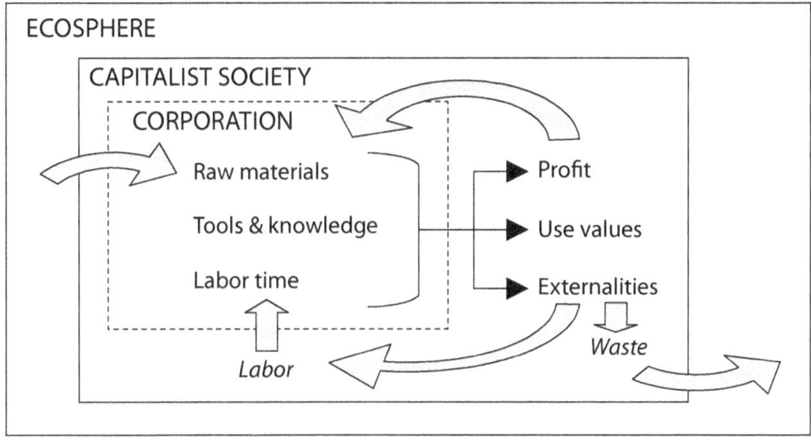

FIGURE 10.3. Social metabolism under corporate capitalism. Source: author.

its proprietary knowledge. It is the corporation that decides what use values will be produced and for what market, what costs will be externalized onto whom, and how waste will be disposed of (or not). Today, the social metabolism between humans and the land is under the control of the corporation. It has stopped serving human needs to serve the corporate need for profit.

CORPORATE MEANING-MAKING AND HUMAN AGENCY

Looking back to my opening remarks, one can wonder why, if the corporation is such a dominant institution, it would be invisible like the water we swim in, or even the fish tank itself. As I explained above, the corporation is not really invisible: we all see it every day. It has been the dominant economic institution—the one mediating the human metabolic relationship with the land—in the Western world since the seventeenth century, and then globally starting at least since the early twentieth century. States entrusted the corporation with the processes involved in satisfying all human needs and wants. From this position of power, it managed to create discourses that justify its existence and make it appear as necessary for society to function. It seems that the corporation has always been with us, will always be with us, and is the best way to organize society and social metabolism. These discourses make the corporation disappear from sight *as a problematic institution*. In its public communications, the corporation presents its interest as the interest of society in general: it provides people with jobs and governments with tax revenue, implying that its activities should be supported and that its profits are legitimate. Canadian sociologist William Carroll and I explored elsewhere how networks of corporate-funded organizations like business associations, think tanks, and foundations create and promote this discourse through the media, education programs, and government lobbying, such that it has now become common sense (Carroll and Sapinski 2010, 2018). On a deeper level, the corporation draws on myths deeply ingrained in Western culture. In these respects, it presents itself as the main actor of human progress: its special power to "innovate" and its capacity to wield "technology" allow it to claim control over the process of social evolution itself, that the myth of progress ties to ever-more complex tools and machines (Sapinski 2015; Sapinski and Surette 2024).

Yet the corporation remains a meso-level legal entity constructed following the historical decisions of elected officials and state managers. It has served up to now as a powerful vehicle to consolidate capitalist class interests. But the future is not written; it is made every day by a multitude of individual and institutional actors within historically constructed structures that constrain and enable their agency. As it is sociologists' role to shed a critical light on everyday life and connect it with the broader institutions and structures in which it takes place (Mills 1959; Smith 1987), I believe it is the task of environmental sociology to address the corporation as the totalitarian institution that it is, and to bring to light the role it plays in connecting humans to their environment. Other institutions and discourses exist that support a future where humans fully control how they live on the land (Atleo 2012; Norgaard 2018; Salleh, Goodman, and Hosseini 2016). By foregrounding the corporation, at the same time that we reproblematize its institutional existence, we also put forth these alternative modes of organizing human life.

REFERENCES

Abraham, Yves-Marie. 2016. "Sortir de l'entreprise-monde." *Possibles* 40 (2): 102–16.

Atleo, E. Richard. 2012. *Principles of Tsawalk: An Indigenous Approach to Global Crisis*. University of British Columbia Press.

Bair, Jennifer, ed. 2009. *Frontiers of Commodity Chain Research*. Stanford University Press.

Bakan, Joel. 2004. *The Corporation: The Pathological Pursuit of Profit and Power*. Viking Canada.

Bell, Michael Meyerfeld, Loka L. Ashwood, Isaac Sohn Leslie, and Laura Hanson Schlachter. 2020. *An Invitation to Environmental Sociology*. 6th ed. Sage.

Brandon, Pepijn. 2015. *War, Capital, and the Dutch State (1588–1795)*. Brill.
Bridge, Gavin. 2008. "Global Production Networks and the Extractive Sector: Governing Resource-Based Development." *Journal of Economic Geography* 8 (3): 389–419.
Campbell, Elizabeth H. 2021. "Corporate Power: The Role of the Global Media in Shaping What We Kow about the Environment." In *Twenty Lessons in Environmental Sociology*, edited by K.A. Gould and T.L. Lewis. Oxford University Press.
Carolan, Michael. 2020. *Society and the Environment: Pragmatic Solutions to Ecological Issues*. 3rd ed. Routledge.
Carroll, William K., ed. 2021. *Regime of Obstruction: How Corporate Power Blocks Energy Democracy*. Athabasca University Press.
Carroll, William K., and J.P. Sapinski. 2010. "The Global Corporate Elite and the Transnational Policy-Planning Network, 1996–2006: A Structural Analysis." *International Sociology* 25 (4): 501–38.
Carroll, William K., and J.P. Sapinski. 2018. *Organizing the 1%: How Corporate Power Works*. Fernwood.
Davoudi, Leonardo, Christopher McKenna, and Rowena Olegario. 2018. "The Historical Role of the Corporation in Society." *Journal of the British Academy* 6 (s1).
Deneault, Alain, and William Sacher. 2012. *Imperial Canada Inc.: Legal Haven of Choice for the World's Mining Industries*. Talonbooks.
Ferdinand, Malcom. 2021. *Decolonial Ecology: Thinking from the Caribbean World*. Wiley.
Gould, Kenneth A., and Tammy L. Lewis, eds. 2021. *Twenty Lessons in Environmental Sociology*. 3rd ed. Oxford University Press.
Hopkins, Terence K., and Immanuel Wallerstein. 1977. "Patterns of Development of the Modern World-System." *Review (Fernand Braudel Center)* 1 (2): 111–45.
Lebbie, Tamba S., Omosehin D. Moyebi, Kwadwo Ansong Asante, Julius Fobil, Marie Noel Brune-Drisse, William A. Suk, Peter D. Sly, Julia Gorman, and David O. Carpenter. 2021. "E-Waste in Africa: A Serious Threat to the Health of Children." *International Journal of Environmental Research and Public Health* 18 (16).
Liboiron, Max, and Josh Lepawsky. 2022. *Discard Studies: Wasting, Systems, and Power*. MIT Press.
Mills, C. Wright. 1959. *The Sociological Imagination*. Oxford University Press.
Norgaard, Kari Marie. 2018. "The Sociological Imagination in a Time of Climate Change." *Global and Planetary Change* 163:171–76.
Oreskes, Naomi, and Erik M. Conway. 2010. *Merchants of Doubt: How a Handful of Scientists Obscured the Truth on Issues from Tobacco Smoke to Global Warming*. Bloomsbury Press.
Pineault, Éric. 2023. *A Social Ecology of Capital*. Pluto Press.
Prechel, Harland N. 2000. *Big Business and the State: Historical Transitions and Corporate Transformations, 1880s-1990s*. SUNY Press.

Salleh, Ariel, James Goodman, and S.A. Hamed Hosseini. 2016. "From Sociological Imagination to 'Ecological Imagination': Another Future Is Possible." In *Environmental Change and the World's Futures: Ecologies, Ontologies and Mythologies*, edited by J.P. Marshall and L.H. Connor. Routledge.

Sapinski, J.P. 2015. "Constructing Climate Capitalism: Corporate Power and the Global Climate Policy-planning Network." Ph.D. diss., University of Victoria.

Sapinski, J.P., and Céline Surette. 2025. "Glyphosate-Based Forest Management in New Brunswick: Regulatory Context, Socio-Environmental Health, and Industry Discourse." In *The Higgs Years. Leading and Dividing New Brunswick*, edited by G. Arsenault. McGill-Queen's University Press.

Savard, Rémi. 1980. "Le sol américain : propriété privée ou Terre-Mère." *Anthropologie et sociétés* 4 (3): 29–44.

Schnaiberg, Allan. 1980. *The Environment: From Surplus to Scarcity*. Oxford University Press.

Sicotte, Diane. 2016. *From Workshop to Waste Magnet: Environmental Inequality in the Philadelphia Region*. Rutgers University Press.

Smith, Dorothy E. 1987. *The Everyday World as Problematic: A Feminist Sociology*. University of Toronto Press.

Solé, Andréu. 2011. "Développement durable ou décroissance. Le point aveugle du débat." In *Développement durable versus décroissance. Débats pour la suite du monde*, edited by Y.-M. Abraham, L. Marion, and H. Philippe. Écosociété.

Stern, Philip J. 2017. "The Corporation in History." In *The Corporation: A Critical, Multi-Disciplinary Handbook*, edited by G. Baars and A. Spicer. Cambridge University Press.

Taylor, Graham D., and Peter A. Baskerville. 1994. *A Concise History of Business in Canada*. Oxford University Press.

Van Apeldoorn, Bastiaan, and Naná de Graaff. 2017. "The Corporation in Political Science." In *The Corporation: A Critical, Multi-Disciplinary Handbook*, edited by G. Baars and A. Spicer. Cambridge University Press.

11. Navigating Technological Futures

Holly Jean Buck

"We are being lied to," declares the first sentence of Silicon Valley venture capitalist Marc Andreessen's *The Techno-Optimist Manifesto*. "We are told that technology takes our jobs, reduces our wages, increases inequality, threatens our health, ruins the environment, degrades our society, corrupts our children, impairs our humanity, threatens our future, and is ever on the verge of ruining everything. We are told to be angry, bitter, and resentful about technology" (Andreessen 2023).

Andreessen, the billionaire who coauthored the first widely used web browser, is convinced that technology is the solution to environmental challenges. In his techno-optimist's world, there are enemies, and the enemies are bad ideas. "Our present society has been subjected to a mass demoralization campaign for six decades—against technology and against life—under varying names like 'existential risk', 'sustainability', 'ESG', 'Sustainable Development Goals', 'social responsibility', 'stakeholder capitalism', 'Precautionary Principle', 'trust and safety', 'tech ethics', 'risk management', 'de-growth', 'the limits of growth.'" These enemies include the ivory tower with its abstract theories and social engineering, according to Andreessen, and "deceleration, de-growth, depopulation—the nihilistic wish, so trendy among our elites, for fewer people, less energy, and more suffering and death" (Andreessen 2023).

By the time this book goes to print, all kinds of people will have used this manifesto as a foil or a straw person, so let me join in. Actually, you could read this manifesto as a sign of the success of those academics and activists who have championed using less energy, or **degrowth**, a concept that describes a planned reduction of energy and resources in a way that reduces inequality and improves human well-being (Hickel 2021) or using less energy. The fact that one of the world's most successful venture capitalists feels the need to take down these ideas, to name them as an enemy, means they are no longer fringe ideas. Rather, some perceived critical mass of people is viewing them as a viable way forward in the face of environmental crisis.

However, this manifesto is also a blinking red indicator of the trap that environmental sociology risks being stuck in—the trap of binary thinking about the role of technology in dealing with the environment. On one hand, you have, well, Andreessen. On the other hand, you have a school of thought suggesting that confronting environmental crises requires not just curtailing economic growth but decreasing energy use, with language around "sufficiency" (cf. Diana Stuart's chapter in this volume) or "scaling down" the human presence (Crist 2019; quoted in Christine Labuski and Shannon Bell's chapter). This all sounds completely sensible on the face of it. But it quickly can lead one toward the anti-humanist strands of environmentalism, with dark political implications in the real world.

Eileen Crist, for example, the scholar quoted above who advocates "scaling down," lead-authored a "Scientists' warning on population," which calls for "a direct, global appeal to all women and men to choose none or at most one child" (Crist et al. 2022). The Green Revolution, Crist also writes, has "underwritten the explosive growth of the human population." While she acknowledges that the existence of nearly half the population is indebted to Green Revolution technologies, she also sees these technologies as bringing disastrous harms, and advocates for a "deep solution" of scaling our population down to around two billion people, who can be fed by food that is "ecologically and ethically produced" (Crist 2022). Of course, Crist, like most writers on population, declares that the pathway toward this solution is female empowerment, not something nefarious. But that doesn't exactly tie a bow around the implication of her argument: that seven billion people should be considered excess beings.

Yet this kind of limits-based thinking fits naturally into environmental sociology, which grew up together with environmentalism. The roots of environmental sociology were born in a moment of technological disappointment. For example, in the introduction to sociologist William Catton's 1980 book *Overshoot*, former Secretary of the Interior Stewart Udall wrote, "The 1960s were not so hospitable to big technology, or to the proponents of the mind-of-man doctrine. Indeed, they were a decade of disillusionment, of broken promise and unexpected developments" (Catton 1980). The Vietnam War, famines, the energy shocks, and more must have been disappointing to those at the time, and these things led scholars and policymakers to declare, as Udall did, "To accept the hard path of belt-tightening and sacrifices, we must first trim back our technological optimism. We need, in short, something we lost in our haste to remake the world: a sense of limits, an awareness of the importance of earth's resources" (Udall 1977). Along with this sense of limits, there was also an emphasis on small and decentralized technologies, typically nonindustrial technologies or "appropriate" technologies. This could be summed up in environmentalist Amory Lovins's influential notion that there are "hard" and "soft" energy paths, with two mutually exclusive policy directions. The "hard" path involves maximization of energy use for economic and social well-being, reliance on "hard" resources and technologies like nuclear power, continuing economic growth, centralized energy production, and market mechanisms. The "soft" path involves reduction in total energy use, intermediate and appropriate technology, and community scale management and decentralization (Lutzenhiser et al. 2002).

Given this foundation, and the lack of any real challenges to it, environmentalism is more or less stuck in a paradigm of instinctively trying to restrict technological development and using the regulatory apparatus to do so. This is a huge problem. It is a huge problem because navigating environmental crises—such as climate change, water scarcity, or waste—is going to require the design and use of a variety of new and existing technologies deployed at planetary scales. We are going to need to build a whole new energy system, and the level of construction, industrial development, and planning that will be required oftentimes contrasts with conventional environmental approaches, practices, and values.

Those who doubt this need to look into the details of what "scaling down" or "welcoming limitations" will really take. Even those scenarios for navigating climate change that assume lower energy demands rely on technological innovations. For example, consider the Low Energy Demand scenario that underlies some of the modeling used in the Intergovernmental Panel on Climate Change assessments. This scenario manages to meet the 1.5°C target by partly featuring the "digitization of daily life," whereby sensors help with energy systems and devices become increasingly smart and interconnected. In addition, according to this scenario the vehicle fleet must be halved by mid-century enabled by ridesharing platforms and autonomous vehicles (Grubler et al. 2018). Another study solves for 1.5°C by leaning on a technological breakthrough in cultured meat, with 80 percent of meat and eggs replaced by cultivated meat by mid-century, and hydrogen-based air travel, accompanied by restrictions, such as two appliances per household (Van Vuuren et al. 2018). In short, the path toward limiting climate change while also ensuring that a growing global population has access to sufficient energy for a decent life means that environmentalists must embrace at least *some* sort of technological development and deployment.

The need today is for environmentalism to guide development of all kinds of technologies and practices, from the personal to the industrial. However, despite two decades of critique (e.g., Nordhaus and Shellenberger 2004), many popular environmentalist responses to climate change still center on having less of an individual impact. The message, in other words, is that the answer is not in technology but in your behavior, and that these are separate and oppositional avenues of climate action.

Right now, we seem to be stuck in a caricature world with two opposing schools of thought, whether those are characterized as technological optimism and pessimism or ecomodernism and degrowth. Notably, tech-media platforms profit from the polarization between these views of techno-optimism and limitation. Environmental sociology as a field risks getting caught up in this struggle, in part because of the incentive structures within academia. Citations are essentially an attention economy metric, and with citations and publications as key metrics for hiring and promotion, the incentive is to get noticed. Early career scholars who may be in precarious jobs face the pressures of personal branding and attention economy competition more than their tenured colleagues (al-Gharbi, 2024) and thus are especially vulnerable to having to participate in polarized debates around environment and technology. This is not good for the field or the planet.

In the rest of this chapter, I'm going to talk a bit more about what's at stake for environmental sociology now, given this situation, and then suggest some ways forward for the field in terms of how it studies technology and technological practices.

I'll begin with the simple contention that both environmentalism and sociology are coming up short in the face of the climate crisis. At a moment in which we need to act rapidly in the material world, both the cultural movement of environmentalism and the academic field of sociology are limited in their ability to contribute the intellectual tools needed. I've discussed above the long history of the drawbacks of environmentalism in general, but what about sociology in particular?

WHERE IS ENVIRONMENTAL SOCIOLOGY IN DEVELOPING NEW TECHNOLOGIES?

In previous decades, sociology's discussion of the environment and technology centered on risk. Risk has conventionally been defined in terms of the existence of a hazard—and the probability of that hazard having a negative impact. **Technological risk**, then, is the probability and magnitude of adverse effects of technological hazards on human health and safety and the environment (Dietz, Frey, and Morrison 2002). The 1970s and 1980s were a time of highly complex environmental problems like the ozone hole and acid rain, and this period was also dominated by novel technological disasters like Chernobyl and Bhopal. These emergent "mega-hazards" pushed risk itself to be a more central concept in the field of environmental sociology. By the 1990s, globalization had entered the picture. Research focused on how globalization intersects these risks, as well as how knowledge, claims, and policies are influenced by perceptions, culture, and politics, given that many of the new hazards are invisible (Dunlap, Michelson, and Stalker 2002).

Perhaps the most influential figure in these conversations was Ulrich Beck, who, with Anthony Giddens, developed the now well-known concept of "**risk society**," one grounded in the observation that in modern times, the dominant class is not just based on industrial ownership (Beck 1992, Giddens 1998). Rather, the dominant class controls technological knowledge itself—which means that antagonism and class struggle have moved beyond the industrial working class to the middle and professional-managerial classes who are now undertaking these struggles through "scientized" discourses of risk and ecology (Buttel and Humphrey 2002). It doesn't mean, however, that society is necessarily more dangerous and hazardous than in, say, the Middle Ages—it means that there was no similar notion of risk in these earlier eras since dangers are experienced as a given. The idea of risk, as Giddens (1999) explains, is "bound up with the aspiration to control and particularly with the idea of controlling the future" (3). These discussions of risk were salient in that a whole field of risk analysis and assessment emerged and became part of global governance.

Risk assessment is a practice that includes understanding the risk context and estimating the likelihood and severity of the consequences of hazards; risk evaluation involves determining the acceptability of identified risks; and risk management involves steps to mitigate unacceptable risks (Gattinger 2023). Formal risk analysis itself is derived from mathematical projections of expected rates of mortality and morbidity, based on known or estimated relationships between degrees of exposure. While this appears highly technical, it should be understood as an exercise of political power, a social act that can be informed by biases as well as institutional practices (Dietz et al. 2002). Pointing this out does not imply that our risk assessment infrastructure is useless; rather, it shows that our risk assessment infrastructure needs to be understood as the social and political battleground that it is. This makes it an important site of applied environmental sociology.

Since the 1960s, many US agencies, including the Environmental Protection Agency, the Nuclear Regulatory Commission, the Occupational Safety and Health Administration, and the Food and Drug Administration, have emerged to regulate risk. The US risk policy system includes not just these executive branch agencies but a tapestry of other actors, from Congress to law and consulting firms, from corporations and industry associations to environmental organizations, universities, labor, and local governments. Lower-income countries may have constrained capacity to assess and manage technological risks, and technological hazards transcend national boundaries. There are many organizations and treaties that help to manage risks, including the UNEP, the World Health Organization, the IAEA, the Basel Convention on hazardous waste, the Montreal Ozone Protocols, and more (Dietz et al. 2002).

Today, though, assessment of risk is probably more complicated than in the past, owing to the fragmented media landscape that empowers radically different worldviews about environmental risks, whether those be potential harms from offshore wind turbines or management of hydrogen infrastructure. It's almost as if policymakers might want input from sociologists to figure out the way forward—but no one is really turning to sociologists. Sociology could be helpful in order to think about how to enact both massive social and technological transformation, but the discipline somehow doesn't seem to have a clear seat at the table when it comes to implementing federal or global climate policy. Why not?

Given that society needs some sorts of technologies to get us through multiple environmental crises, the question for environmental sociology now is not whether we should advocate for "more" technology or "less" technology, but rather *which* technologies? Who invents them, who can access them, and who benefits from them? How are they scaled? Who bears the externalities or harms?

Environmental sociology *is* asking these questions—but there is little indication that environmental sociologists are doing so in an applied way that can move the creation of new clean industries or practices forward. There are a number of potential reasons for this—and they impact sociology's aforementioned lack of a seat at the table with policy implementation, too. One is that technology has fallen off the radar in environmental sociology. Look no further than the fact that the American Sociological Association's (ASA) section on environmental sociology was once called the "environment and technology" section, but section membership decided to change the name several years ago to just "environmental sociology" because studies on technology had fallen out of favor. Another is that contemporary environmental sociology is interested in progressive activism; as indicated in the introduction, scholars are often concerned with making active connections to a wider variety of social movements. This priority brings a constraint, though, at times when those movements are more interested in restricting development rather than building new industries. Scholars with activist backgrounds may also be hesitant about working with the state, which, in an era of state-deployed clean energy funding, may also be a limitation. In striving to build visible and meaningful alliances with social movements, the field can get aligned with one side of an ideological binary about the role of technology in dealing with environmental challenges. And so, when policymakers, think tanks, or companies start thinking about how to deploy new technologies, they aren't bothering to consult with a group of people whose cards, as technology skeptics, already appear to be on the table.

Part of the irony of the current moment is that sociology has been instrumental in learning to see beyond **technological determinism**, and understand that technology is socially constructed. The environmentalists who write off "false solutions" or "technofixes" tend to essentialize technology. Yet sociology of decades past said there was more to the story: that technology was shaped by humans and their social and political choices. Technology, as Andrew Feenberg (1999) puts it, isn't some "autonomous force separate from society, a kind of second nature impinging on social life from the alien realm of reason in which science too finds its source" (vii). Rather, "technology is the medium of daily life in modern societies," and seeing the technical and the social as separate domains is a trap, Feenberg indicates. His project is to put forth an anti-essentialist philosophy of technology, meaning that technology doesn't just reduce everything to functions and raw materials and sweep away human meaning. To borrow Feenberg's (1999) example, a house is a system with a bunch of gadgets in it, with electrical, communications systems, plumbing, heating, mechanized tech that built it, and so on (and soon, we might add, maybe a device for putting power back into the grid)—but it is still a rich and meaningful life environment (xi). And those dimensions are intertwined—the coziness of the house is also connected with its heating system.

On one hand, sociological theory from the past few decades points us toward insights like the following: formal risk assessment is socially constructed; bureaucracies who deal with this have practices for evading responsibility; publics are going to perceive risks based on their worldviews and prior attitudes and media signals rather than these formal assessments anyway. There is a way in which the collective inheritance of environmental social science on risk and technology could be read as paralyzing.

On the other hand, we need to develop technologies in the public interest in order to deal with climate change, and reproducing the critique of the past few decades for every new technology or project is just not going to cut it.

Environmental sociology is thus at a crossroads. Repeating critique will not be sufficient, because large language models trained on critical theory will be able to produce critique of new innovations for us—plus, our students are hungry to actually do things that they feel are making a difference. Often, what we offer instead is coproducing research with and in service of community groups to produce the data they need to successfully fight technological projects that they do not deem to be in their interest. This is an important role for the field, but ultimately it will not be enough, since it is fundamentally a reactive role. Contemporary environmental sociology has more or less left engagement with technology-shaping practices to architects, designers, planners, entrepreneurs, and the practitioners of other future-oriented disciplines. It is missing an opportunity.

APPLIED ENVIRONMENTAL SOCIOLOGY FOR REMAKING THE WORLD

The time is right for a type of applied environmental sociology that maintains the critical inheritance of the discipline while also guiding new technologies to be developed and deployed in the public interest. Other authors in this book have called for an "implementation environmental sociology" research agenda (Houser, chapter 2, this volume) or an "engaged public sociology" (Cordner, chapter 4, this volume); this is very much aligned. In what follows, I suggest three directions to take when it comes to technology.

Anticipating the Social Impacts and Justice Implications of New Technologies

Social impact assessment (SIA) is a field that emerged at the same time as technology assessment, in the 1960s and 1970s, but it has not been as fully institutionalized, as, for example, environmental impact assessment. In the United States, social impact assessment is a part of the National Environmental Policy Act (NEPA), and SIA has focused on several things over the years, such as assessing construction projects, economic opportunities from projects, boomtown disruptions, resource use plans, policies and programs, and international development projects (Finterbusch and

Freudenburg 2002). Social impact assessment that does a better job with examining distribution of risks and benefits may also provide an evidence base for the negotiation of compensation; at the same time, social science reminds us that these assessments do not just produce "the facts" but are socially constructed processes themselves (Walker 2010).

Assessing the socio-ecological impacts of new technologies is not just a matter of project-scale analysis. Developers, researchers, and policymakers also need to understand the implications of new technologies at scale. The literature on anticipatory governance and responsible research and innovation is relevant here (Guston 2014; Owen et al. 2021), and an applied environmental sociology could draw from this as well as interdisciplinary methodologies, such as scenarios, foresight (Selin 2008), and more.

One reason why social impact assessment takes on new relevance is the efforts underway to institutionalize environmental justice-oriented assessment. For example, in the United States, the states of New York and New Jersey passed legislation expanding existing environmental review to require environmental justice considerations and cumulative impact assessments for permitting projects that might affect the environment, meaning that agencies have to consider the cumulative impact of their actions on disadvantaged communities, not just consider each potential activity separately. Owing to structural racism, polluting facilities are disproportionately sited in communities of color; therefore, cumulative impact analysis could be an important tool in ensuring that the new system is built without increasing burdens—assuming that there are mandates that analysis is used in making siting decisions. At a time when a new energy system needs to be built out—as well as new waste handling and recycling facilities, clean manufacturing, new transportation networks, and more—this could impact where and how new environmental technologies are deployed, and how their social impacts are considered. Helping society continue to make environmental justice a part of social impact assessment could be an important task for environmental sociology today.

Using Sociological Insights and Methods to Foster Public Participation

People want to be involved in decisions that impact them and their environment. But if laypersons view risk differently than experts do, should they be involved in risk management? As one handbook (Gattinger 2023) summarizes, "Will involving the public lead to stronger or weaker risk decisions? Will it compromise evidence-based decision-making for risk or help to inform it? Will it amplify risk controversies or help to resolve them?

Will it politicize risk decision-making or foster democratic accountability? Will it build or erode public trust in risk decisions?"

Within social science, there have been different schools of thought about public participation in environmental risk assessment and environmental management. Psychologists have studied how people systematically under- or overestimate risks, identifying mechanisms like the availability heuristic, where an event that is vivid and easy to recall strongly influences the perceived probability of the event (Dietz et al. 2002; Tversky and Kahneman, 1982). Some might say, if the public is shown to often misjudge the risks, decisions about things like how to regulate nuclear power or GMOs should be left to the experts, who understand the technical probabilities of adverse effects. Others might say that the experts have their own cognitive biases, and it's not all about technical risk anyway, and so laypersons should be involved in assessment. In general, the social science literature assesses public involvement positively, with many examples of how participation strengthens risk decisions—that it is fundamental to democratic accountability, and that it can foster trust (Gattinger 2023). There's also a normative demand coming from many communities for procedural justice, or for having power and involvement in how environmental decisions are made. From an environmental justice perspective, it might be obvious that people who live near a facility, including those that deal with environmental risks, have the right to be involved in decisions about it.

Ideally, environmental sociology might be in a position to help different actors to understand each other, including power dynamics, views on risks, and values, and to provide tools or practices to work through risk controversies. Right now, there is often not any clear actor who is tasked with "participation" or "engagement," outside regulatory processes that might require a public hearing or something to that effect. The academy—particularly, public institutions that have the community and public as part of their mission—could step up.

This isn't an easy place for sociologists to be in, because of tensions around working with the state, which may be leading participatory processes. Some environmental justice advocates and scholars have observed that participation in state-led processes may not be enough to foster participation. Writing from a critical environmental justice perspective, David Pellow (2020) observes that a vision of change that looks to the state to accommodate demands via legislation and reforms may leave intact the power structures that produced environmental justices and, by collaborating with the state, risk reinforcing its legitimacy. "While I agree that justice via procedural inclusion and recognition can be important to the future of the EJ movement and to any community's efforts to create change, the reality is that it is often a step toward a more sophisticated effort at differential inclusion, co-optation, displacement of movement goals, diffusion of grassroots energy, assimilation, and a strengthening of existing power relations," Pellow writes. Similarly, Laura Pulido and colleagues offer a solidarity-based critique of the EJ movement, suggesting that it has had only minimal success in improving the environments of vulnerable populations aside from blocking new projects, because it relies so much on state regulation. They urge the movement to rethink its attitude toward the state and refuse to participate in regulatory charades: "The EJ movement should take a page from Black Lives Matter. *It's not about being respectable, acknowledged, and included. It's about raising hell for both polluters and the agencies that protect them.* Given the planetary crisis we are facing, we need a radicalized EJ movement more than ever" (Pulido et al. 2016; italics in original). Neither do sociologists want to spend time participating in regulatory charades.

On the other hand, there is the danger that in rejecting the potential of the state to govern risk ends up strengthening neoliberalism, with the state's legitimacy now under attack from various fronts. If elected officials cannot make decisions about environmental risk, is it more legitimate to have unelected groups of self-appointed community leaders pushing for their views of what environmental risks are appropriate? How does a system like this manage tradeoffs between local harms and national or global benefits? Most likely, it doesn't. This whole area of participation and engagement is one area that could use labor-hours and attention from a justice-oriented applied environmental sociology. Sociology—science and technology studies in particular—has been key to developing conceptual language around "responsible innovation" or "public interest technology," and public participation in both decision-making and innovation is fundamental in both of these. Applied environmental sociology could help public institutions further operationalize these concepts so that technology can be better guided by public values and priorities.

Developing New Theory and Rethinking Ecological Modernization

Today, "ecomodernism" is practically used as a casual epithet in many environmental sociological circles, but ecological modernization theory was an important development within environmental sociology. Ecological modernization theory placed an emphasis on improvement, and expanded the field's traditional focus beyond degradation (Dunlap et al 2000). "**Ecological modernization**" is a term that can refer to (1) a new concept bringing theoretical contributions to environmental sociology, (2) a concrete program of environmental policy, and (3) a literature grounded in social science studies that analyzes environmental policies (Lenzi, 2022; Mol 1995). As a literature, it examines ecological reform, and how companies and governments are responding to the ecological crisis through sustainability initiatives, such as energy efficiency or changing supply chains. The idea is that growth can be decoupled from resource extraction and waste production: economic growth could continue without environmental impacts. Moreover, as societies develop materially, individuals and groups become more "green."

Other theories covered in this volume, such as treadmill theory or degrowth (see chapter 2), contest the ability of capitalism to reform itself. Critics of ecological modernization theory also point to the Jevons paradox, which is when increases in efficiency actually increase demand for a resource. Key critiques of ecological modernization theory originating in environmental sociology include Richard York and Eugene A. Rosa (2003), who make four key challenges to the theory. The first has to do with institutional efficacy. They argue that ecological modernization theory needs to demonstrate that the institutions of late modernity are effective at addressing ecological challenges, and evidence here is lacking. Second, they point out that case studies are an ineffective way to prove that societies are actually moving in the direction of sustainability. Third, they put forth a critique around units of analysis: focusing on individual organizations and sectors, they state, is not enough to illuminate what is happening on a macro level. Fourth, they critique the singular focus on efficiency; improvements in efficiency may not outweigh the expansion of production that comes with modernity. Given these weaknesses of ecological modernization theory, York, Rosa, and Dietz (2003) offer an alternative perspective—namely, the political economy perspective, which says that environmental exploitation is driven by the structure of market economies, institutions, and the commitment to growth that is inherent in capitalist production systems. Moreover, world-systems theory, which applies the political economy perspective at a global scale, points to the ways in which analyses offering evidence of decoupling or dematerialization in some countries are failing to see the full picture, since countries in the world's core are shifting the negative environmental impacts from their growth to the world's peripheries—for example, by exporting waste or by outsourcing production.

This all leads some theorists to recommend degrowth, often defined along the lines of being an equitable downscaling of production that increases human well-being and enhances ecological conditions at the local and global level, in both the short and the long term (Koch, Buch-Hansen, and Fritz 2017). Degrowth has long critiqued technological fixes, and degrowth scholars have examined the prospects of technologies that are convivial, small-scale, decentralized, and low-tech/low-energy intensive, as well as locally manufactured (Kallis et al. 2018).

Some critics of degrowth say that growth is not the cause of environmental challenges, but that capitalism is. Others point out that degrowth might be a valuable strategy when focusing on the world's richest billion people, but climate responses need to account for the other billions of people whose basic material needs are not yet met. As Jonathan Symons (2019) writes, "Were all of humanity to converge on a standard of living typical of middle-income countries—replete with mattresses for sleeping, cooking stoves, reliable electricity and refrigerators, but excluding Western luxuries like long-distance travel, private washing machines or cars—this would still entail a massive increase in consumption" (50). So, even though degrowth promoters explain that it is only high-income, disproportionate-using consumers and nations that need to degrow (Hickel 2021), on a global level, energy use is set to increase because of the lack of access to clean cooking fuels and electricity that millions currently face. Our solutions, in other words, must embrace new technological development and distribution, even if we are also focusing on how to curb consumption of the highest-income consumers.

Ecological modernization theorists also point out that even with increasing attention to the dangers of, for example, chemicals, there is no massive movement away from a lifestyle dependent on chemicals or the dismantling of chemical production; rather, the emphasis from environmental groups has been to make a sustainable chemical industry (Mol 2003). The evidence that degrowth can be a popularly supported political project is lacking. In fact, the politics of scaling down beef or private transportation (as suggested by Hickel, 2021) are likely to open up a backlash that will end up falling not on the academics promoting the idea but on minoritized communities. For example, one risk of introducing degrowth includes opening the door to severe austerity measures in which essential public services are cut in the name of degrowing the economy. Critics also question how Eurocentric the project can be: while global perception of the idea has not been studied, a recent survey of climate policy researchers indicated that while 36 percent of EU climate policy researchers supported degrowth, only 9.6 percent in non-OECD countries and 6.3 percent in BRICS countries support degrowth (King et al. 2023). This indicates that the scholarly enthusiasm may not be shared globally even within other academic and professional circles.

Environmental sociologists should rethink their approach to all this. Ecomodernism can draw on scientific ingenuity and is more likely to impact the world outside the ivory tower than strategies that reject technologies out of hand (Symons 2019). The differing theories about what is desirable versus what is possible make ecological modernization and degrowth theories rich ground for empirical environmental sociology.

There are a number of theoretical and empirical explorations that would be relevant here. First, what are the real-world implications of and possibilities for degrowth? A justice-centered framework could unpack the direct and indirect effects of degrowth policies and choices not to invest in or develop particular technologies, especially for vulnerable peoples in both overburdened communities in the Global North and the Global South. Second, are our theories of change correct? The implicit theory of change in much environmental sociological scholarship is that a coalition of movements from below will form broader coalitions and force reform from the bottom up. When and how this actually happens deserves even more study, especially in the wake of the failed social uprisings across the Middle East and North Africa in the early 2010s (see Bevins, 2023). Social movements and collective action have been a topic of scholarship for decades, from work examining how the values of social movements have entered entrepreneurship, design, and industrial innovation (Hess, 2007) to when and how direct action is most successful (Fisher, 2024). When it comes to technology in particular, continued research of this sort could help us understand whether and how social movements can direct technological development to minimize harms and maximize environmental and social benefits. Third, how does environmental reform happen? There is a lot more empirically informed theoretical work to be done. A more technologically curious environmental sociology could bring new rigor and depth to these conversations.

CONCLUSION

If sociology does not take up the task of applying social science methods to these important questions, as a society, we will be stuck at an impasse wherein different people shout their different arguments about how to deal with technology and environmental change. We will fail the moment and be unable to plan, let alone enact, policies that can help us to better navigate our collective technological futures. We risk failing to confront climate change. Our fear of technology and weak trust in our institutions, on one side, and a lack of reckoning with historical and current environmental justice, on the other, will leave us unable to do much at all. This is not to claim that social science can solve this impasse; the discipline, however, can contribute insights through empirical research. Moreover, if we simply join the conversation as activists promoting our favorite ideology without some amount of objectivity and empirical methods, we risk continuing insular conversations that don't help society, and we will fail to attract students—who already question the worth of a sociology degree. This would be a shame, because governments, NGOs, and companies are all seeking people who know how to do public engagement or environmental justice assessment, and who can anticipate the social impacts of emerging technologies. An applied environmental sociology that centers justice while maintaining critical thinking about technology is possible, if we choose to create it.

REFERENCES

Al-Gharbi, Musa. 2024. *We Have Never Been Woke: The Cultural Contradictions of a New Elite.* Princeton University Press.

Andreessen, Marc. 2023. "The Techno-Optimist Manifesto," Andreessen Horowitz. https://a16z.com/the-techno-optimist-manifesto/.

Beck, Ulrich. 1992. Risk Society: Towards a New Modernity. Sage.

Bevins, Vincent. 2023. If We Burn: The Mass Protest Decade and the Missing Revolution. PublicAffairs.

Catton, William R. 1980. *Overshoot: The Ecological Basis of Revolutionary Change*. University of Illinois Press.

Crist, Eileen. 2022. "Less Is More." Contribution to GTI Forum: The Population Debate Revisited. Great Transition Initiative. https://greattransition.org/gti-forum/population-crist.

Crist, Eileen, William J. Ripple, Paul R. Ehrlich, William E. Rees, and Christopher Wolf. 2022. "Scientists' Warning on Population." *Science of The Total Environment* 845:157166.

Dietz, Thomas, R. Scott Frey, and Denton E. Morrison. 2002. "Risk, Technology, and Society." In Dunlap and Michelson, *Handbook of Environmental Sociology*.

Dunlap, Riley E., and William Michelson, eds. *Handbook of Environmental Sociology*. Greenwood Press.

Dunlap, Riley E., William Michelson, and Glenn Stalker. 2002. "Environmental Sociology: An Introduction." In Dunlap and Michelson, *Handbook of Environmental Sociology*.

Feenberg, Andrew. 1999. *Questioning Technology*. Routledge.

Finterbusch, Kurt, and William R. Freudenburg. 2002. "Social Impact Assessment and Technology Assessment." In Dunlap and Michelson, *Handbook of Environmental Sociology*.

Fisher, Dana. 2024. *Saving Ourselves: From Climate Shocks to Climate Action.* Columbia University Press.

Gattinger, Monica, ed. 2023. *Democratizing Risk Governance: Bridging Science, Expertise, Deliberation and Public Values*. Springer International.

Giddens, Anthony. 1998. "Risk Society: The Context of British Politics." In *The Politics of Risk Society*, edited by J. Franklin. Polity.

Giddens, Anthony. 1999. "Risk and Responsibility." *Modern Law Review* 62 (2): 1–10.

Grubler, Arnulf, Charlie Wilson, Nuno Bento, Benigna Boza-Kiss, Volker Krey, David L. McCollum, Narasimha D. Rao, et al.. 2018. "A Low Energy Demand Scenario for Meeting the 1.5 °C Target and Sustainable Development Goals without Negative Emission Technologies." *Nature Energy* 3 (6): 515–27.

Guston, David H. 2014. "Understanding 'Anticipatory Governance.'" *Social Studies of Science* 44 (2): 218–42.

Hess, David. 2007. *Alternative Pathways in Science and Industry: Activism, Innovation and the Environment in an Era of Globalization*. MIT Press.

Hickel, Jason. 2021. "What Does Degrowth Mean? A Few Points of Clarification." *Globalizations* 18 (7): 1105–11.

Kallis, Giorgos, Vasilis Kostakis, Steffen Lange, Barbara Muraca, Susan Paulson, and Matthias Schmelzer. 2018. "Research On Degrowth." *Annual Review of Environment and Resources* 43 (1): 291–316.

King, Lewis, Ivan Savin, and Stefan Drews. 2023. "Shades of Green Growth Scepticism Among Climate Policy Researchers." *SSRN Electronic Journal*.

Koch, Max, Hubert Buch-Hansen, and Martin Fritz. 2017. "Shifting Priorities in Degrowth Research: An Argument for the Centrality of Human Needs." *Ecological Economics* 138:74–81.

Kroll-Smith, Steve, Stephen R. Couch, and Adeline G. Levine. 2002. "Technological Hazards and Disasters." In Dunlap and Michelson, *Handbook of Environmental Sociology*.

Lutzenhiser, Loren, Craig K. Harris, and Marvin E. Olsen. 2002. In Dunlap and Michelson, *Handbook of Environmental Sociology*.

Mol, Arthur P.J. 2003. "The Environmental Transformation of the Modern Order." In *Modernity and Technology*, edited by T.J. Misa, P. Brey, and A. Feenberg. MIT Press.

Nordhaus, Ted, and Michael Shellenberger. 2007. *Breakthrough: From the Death of Environmentalism to the Politics of Possibility*. Houghton Mifflin Harcourt.

Owen, Richard, René Von Schomberg, and Phil Macnaghten. 2021. "An Unfinished Journey? Reflections on a Decade of Responsible Research and Innovation." *Journal of Responsible Innovation* 8 (2): 217–33.

Pellow, David N. 2020. "Critical Environmental Justice Studies." In *Environmental Justice*, edited by B. Coolsaet. Routledge.

Pulido, Laura, Ellen Kohl, and Nicole-Marie Cotton. 2016. "State Regulation and Environmental Justice: The Need for Strategy Reassessment." *Capitalism Nature Socialism* 27 (2): 12–31.

Selin, Cynthia. 2008. "The Sociology of the Future: Tracing Stories of Technology and Time." *Sociology Compass* 2 (6): 1878–95.

Symons, Jonathan. 2019. *Ecomodernism: Technology, Politics and the Climate Crisis*. Polity.

Udall, Stewart L. 1977. "The Failed American Dream." *The Washington Post*, June 11.

Van Vuuren, Detlef P., Elke Stehfest, David E. H. J. Gernaat, Maarten Van Den Berg, David L. Bijl, Harmen Sytze De Boer, Vassilis Daioglou, et al. 2018. "Alternative Pathways to the 1.5°C Target Reduce the Need for Negative Emission Technologies." *Nature Climate Change* 8 (5): 391–97.

Walker, Gordon. 2010. "Environmental Justice, Impact Assessment and the Politics of Knowledge: The Implications of Assessing the Social Distribution of Environmental Outcomes." *Environmental Impact Assessment Review* 30 (5): 312–18.

York, Richard, and Eugene A. Rosa. 2003. "Key Challenges to Ecological Modernization Theory." *Organization & Environment* 16 (3): 273–88.

York, Richard, Eugene A. Rosa, and Thomas Dietz. 2003. "Footprints on the Earth: The Environmental Consequences of Modernity." *American Sociological Review* 68:279–300.

12. The Future of Food

Amalia Leguizamón

There is no denying our food system is in crisis. Conventional agriculture, propelled by the agricultural innovations of the 1970s Green Revolution, has led to unprecedented global food production (Patel 2013). However, despite these advancements, hunger continues to afflict one in every ten people worldwide (UN News 2023). The expansion of mechanized, large-scale, corporate-driven agriculture has wrought havoc on society and the environment. Rural communities are contending with a burgeoning health crisis, exposed as they are to the toxic fallout of agrochemical spraying (Harrison 2011; Leguizamón 2020). The expansion of agricultural frontiers encroaches on peasant and Indigenous territories, instigating violent conflicts and dispossession. The toll on natural resources is alarming, resulting from contamination and depletion of water and soil and rampant deforestation. The global food system is responsible for a quarter of greenhouse gas emissions contributing to climate change (Lynch et al. 2021). To confront this crisis, we must rethink food provisioning on the local, regional, and world scales.

The future of food is already underway and is taking two divergent paths. One is the path of ecological modernization, advanced by corporations with cutting-edge technologies. The other is the agroecological movement, presented by Indigenous and peasant movements. As the chapters in this final section demonstrate, not all futures are equally possible—contemporary power structures, ideologies, and institutions constrain what paths we can take. Yet this chapter takes a stand for the agroecological way as a paradigm shift. I position myself alongside Latin American critical social scientists who make a theoretical and epistemological commitment to question and affirm the role of academics and higher education institutions in the social construction of knowledge (Lander 2000). As such, and in line with the collective efforts presented in this volume, I present the agroecological movement as a future for social and ecological justice.

This chapter begins by explaining why food needs a future. The following section introduces the reader to how we got to where we are: the globalized, corporate food system. This will not be exhaustive, as it is not possible to summarize existing literature; nor is that this volume's goal. I do, however, provide citations that will guide the novice reader on the sociology of food and agriculture if they want to read more about this topic. Instead, I take a bird's-eye view looking over the long duration of the spread of capitalism through the expansion of agrarian frontiers. I highlight unequal power dynamics in decision-making over the control of natural resources and technological innovation, as well as defining the narratives, ideologies, and epistemological frameworks that sustain and legitimize them, in particular, a binary, hierarchic distinction that leads to the social control and conquering of nature. The following sections present the two paths for the future of food: one, of food grown in labs by robots, and two, of the agroecological movement. Finally, the conclusion summarizes and revisits the call for agroecology as a socially just and environmentally sustainable alternative to the corporate food system.

HOW WE GOT WHERE WE ARE

We need to take a world-historical view to understand the global food crisis. The transition from local farming cultures to large-scale commercial agriculture has its origins in the colonial division of labor (McMichael and Weber 2021). During the seventeenth and eighteenth centuries, European colonial powers (such as Spain, Portugal, and Britain) set a pattern of extraction across their colonies in the Americas, Asia, and Africa. The colonies specialized in extracting raw materials and producing primary goods, such as silver and gold, spices, cacao, sugar, and cotton, which were exported and exchanged for European manufactured goods and African slaves. In *Sweetness and Power*, Sidney Mintz (1986) traces the environmental history of sugar from a rare luxury to an everyday staple. He shows how the so-called "triangle of trade" fed British factory workers while providing an overseas market for their manufactured goods, creating a system of unequal ecological exchange, where the land and labor of the New World subsidized the Industrial Revolution, the rise of capitalism, and nineteenth-century British hegemony.

The imperial thirst for sugar, spices, and profits decimated people and places worldwide. Raj Patel and Jason Moore (2017) trace the historical roots of the climate crisis to Madeira, a small African island in the Atlantic Ocean. In the 1500s, the Portuguese deforested the island, first for lumber and then for wood, to fuel a new system for sugar production and distribution. Amitav Ghosh (2022) traces it to the Banda Islands in the Indian Ocean, where, in the 1600s, the Dutch burned down towns and massacred their inhabitants to grow nutmeg as a commodity for the spice trade. These stories and their authors teach us to look for the historical roots of contemporary socio-ecological disruption in colonialism and the rise of capitalism. It is at this time that a pattern of extraction for capital accumulation through the expansion of frontiers is established (Patel and Moore 2017). There also arises a particular way of knowing and being in the world: A modern, Eurocentric mentality that established a hierarchic, dichotomic distinction between nature and society, which continues to justify and promote the conquest and domination of non-European peoples, cultures, and terrains (Ghosh 2022; Patel and Moore 2017).

These are the seeds of the modern commercial food economy. The colonies, later the Third World and presently the Global South, were reduced to specialized export monocultures (exemplified by sugar, bananas, and soybeans) to meet foreign needs. Plantation monocultures established agro-industrial logics—large in scale, reliant on constant investments in technological innovations to increase efficiency and productivity from labor and nature, with a relentless pursuit of homogenous, reliable (controllable) harvests—all of which aimed to grow food as a commodity for profit. It first happened in the sugar plantations in the British West Indies, with its innovative combination of field and factory under one authority, the organization of skilled and unskilled labor force around the productivity of the plantation, and a strict, time-conscious scheduling (Mintz 1986).

The banana plantation in twentieth-century Central America intensified these trends, becoming larger in scale, with a more complex division of labor, and operating at a quicker pace while incorporating substantial innovations: large, transnational corporations monopolizing land and transport routes (the origins of vertical integration) and an inclination to address the ecological problems resulting from industrial monocultures with further rounds of technological innovation, including fossil fuels and chemicals. In *Banana Cultures*, John Soluri (2005) presents the environmental history of the expansion of banana monocultures in Honduras. Bananas were grown for mass production and importation to the United States by a handful of companies that, with time and a series of mergers, would first become the United Fruit Company and later, Chiquita, currently one of the largest banana exporters in the world. In Honduras, fruit companies' engineers drained swamplands and cut down forests in the name of modernity and progress to make room for banana export farms. In an effort to control plant disease epidemics, fruit companies funded expensive breeding programs to develop pest-resistant varieties of bananas. When those efforts failed, scientists depended on chemical treatments; as a last resort, they abandoned the infested farm and expanded the agrarian frontier.

In the twenty-first century, soybean monocultures have perfected agro-industrial dynamics. Soybeans are the darling crop of the **corporate food regime**, the dominant form of contemporary global food provisioning (McMichael 2013).[1] Transnational corporations source soybeans interchangeably from Argentina, Brazil, or Paraguay to feed intensive animal farming operations worldwide and increasingly in Asia (Oliveira and Hecht 2016). I study soybean production in Argentina, the third-largest grower and exporter of soybeans worldwide (Leguizamón 2014, 2020). During my fieldwork, soybean farmers told me the crops they grow are not to feed their families. They grow soybeans as a commodity for profit in fully mechanized large-scale farms. Soybean producers employ a technological package of genetically modified (GM) seeds, no-tilling machinery, and glyphosate herbicide to cut labor and input costs, simplify production practices, and increase profitability. Proponents of GM crops argue technological innovation will solve global hunger, poverty, and climate change. Yet, in practice, the expansion of the GM soybean frontier in Argentina has brought violent dispossession and displacement, deforestation, and illness and death owing to exposure to toxic agrochemicals.

Discourses and epistemologies are embedded in and legitimize the historical, global expansion of agrarian frontiers to establish export monocultures. Vandana Shiva (1993) presents "monocultures of the mind" as a metaphor to critique the dominant, narrow, and homogenizing ways of knowledge-and technology-transfer systems, which are promoted by corporations and institutions of the Global North over the agricultural systems and peoples of the Global South. Boaventura da Souza Santos (2014) speaks of "monocultures of knowledge" to critique the dominant Western epistemological framework, which tends to prioritize and valorize formal, instrumental, and economic rationality over other knowledges. Moving forward, I analyze the two emergent paths toward the future of food as they reproduce or challenge capitalist power dynamics over resource use and technological innovation (the relentless expansion of frontiers for capital accumulation) and the epistemic narratives that sustain it (the dominant, monoculture mentality that seeks to make nature "efficient" and profitable).

LABS AND ROBOTS

Technological innovation in food production promises to grow more food using fewer resources. As land and water become increasingly scarce and the world population grows, and as farmers and consumers grow concerned about toxic agrochemicals and animal cruelty, one alternative future path for food provisioning is the path of **ecological modernization**. This path aims to address the socio-environmental impacts of agro-industrial production with green technology (Buttel 2000). Let me show you what this looks like.

In a cotton field in Arkansas, a Silicon Valley engineer is testing a robot weeder (Little 2019:88–109). The robot is hitched to the back of a tractor, connected to eight computers, twenty-four live-feed video cameras, and three large tanks filled with agrochemicals. A software engineer perched in the tractor cab follows the live video of the ground beneath the robot. The robot scans the cotton plants and, in a split second, distinguishes the plant from the weed and applies a targeted dose of chemical herbicide. Thus, instead of applying lots of more toxic chemical herbicides—the conventional path of agro-industrial practices—the robot weeder is a form of precision agriculture that aims to lower chemical use, thus reducing the environmental impact and saving farmers money on expensive chemical inputs.

In California, scientists and engineers from UPSIDE Foods and GOOD Meat grow meat in laboratories (NPR 1A 2023). To make cultivated meat, scientists extract cells from an egg or living animal (mostly chicken).[2] The cell line is then placed into a steel tank bioreactor called the "cultivator," where cells are fed with amino acids, vitamins, fats, sugar, and salts. Two to six weeks later, the grown cells are "harvested." The final step is to give this mass of cells the shape, taste, and texture of meat (for example, to make it look and taste like a chicken breast)—a process achieved using 3-D printing, extrusion, and molding technologies. Cultivated meat promises to replace conventional animal farming, which is cruel to animals and harmful to the planet (animal farming releases methane and carbon dioxide in addition to the social and environmental costs of growing commercial crops like corn and soy to feed them).

Meanwhile, in Argentina, soybean producers and agribusiness leaders claim that knowledge holds the key to growing more food to address world hunger and poverty with minimum ecological impact (Leguizamón 2020). They argue that the future of agriculture and sustainable development lies in corporate-sponsored scientific and technological developments, like genetically modified crops and no-tilling machinery, and in novel arrangements for land and labor. The new agricultural paradigm advanced by Los Grobo, one of the largest agribusinesses in Argentina and the world, is known as the "knowledge-based network model." The model is centered on building a network of input and service providers, including landowners, agronomists, contractors, and branch managers. Therefore, instead of owning land and machinery and hiring farm workers, the company operates through land leases and third-party contracting. In the agribusiness paradigm, the value of agriculture thus lies not in the material conditions that make it possible (land, labor, climate) but in the expert knowledge to manage and organize production and in the scientific knowledge to develop new agricultural technologies. As J.P. Sapinski demonstrates in chapter 10, corporations actively create and promote discourses that make their dominance appear natural and necessary. In agriculture, this includes presenting corporate-controlled technological and organizational "knowledge" as the only path to addressing global food challenges, while obscuring how this same corporate control perpetuates hunger and environmental degradation.

These examples share a vision for the future of food provisioning based on the promise of ecological modernization. Corporate leaders and scientific experts advance a model for agriculture that seeks to achieve food security in the most efficient, allegedly sustainable, and profitable manner. They seek to overcome the environmental exhaustion and pollution resulting from the centuries-old expansion of agro-industrial monocultures with yet more technological innovation. The discourse that sustains capitalist accumulation reproduces the monoculture mentality of Western modernity. Agribusiness leaders and experts seek to "feed the world" by making the natural conditions that support food production efficient, measurable, reliable, and profitable. What could be more rational and efficient than growing a chicken from a cell in a bioreactor or, as in Argentina, to farm with technocratic "knowledge"?

The digital technologies of precision agriculture (robot weeders and satellites, computers, and nanotechnology sensors, among others) allow the experts to monitor and manage fields from a distance (Leguizamón 2016, 2020). The distance between those who decide what is grown, the farm, and what is eaten, further increases with the financialization of the food system. We see it in the financialization of commodities (Clapp 2014), land (Fairbairn 2014), and the outpouring of venture capital to finance technological development in food production. For example, UPSIDE Foods raised funds from corporate tycoons Bill Gates and Richard Branson, the world's largest poultry producer, Tyson Foods, and investment companies like SoftBank Group, Norwest, and Temasek.[3]

A future of food characterized by increasing digitalization and financialization of the corporate food regime further exacerbates unequal power dynamics over decision-making and the uneven distribution of costs and benefits. As Holly Buck argues in chapter 11, environmental sociologists' concerns over technological innovations are not about the technologies themselves. The key question is about who controls them and to what ends. This future for food reliant on labs, robots, and venture capital continues to grow what Phil McMichael (2013) calls "food from nowhere" (18). The globalized nature of the food system persists, and the disconnect between food production and consumption increases. Consumers continue to be unaware of where their food comes from, how it is produced, and the social and environmental implications of its production. Food is only available for those who can pay for it. The cotton grown by robots in Arkansas and the soybeans grown with "knowledge" in Argentina continue to be grown as plantation monocultures for export. Lab-grown meat is currently produced on a small scale (and is neither profitable nor efficient), so UPSIDE and GOOD Food CEOs are actively seeking capital to scale up (NPR 1A 2023). Furthermore, not one of the examples I presented has replaced food conventionally grown by the corporate global food system—they have only added new products and technologies. Capital has efficiently expanded into new frontiers to accelerate accumulation.

THE AGROECOLOGICAL MOVEMENT

Agroecology is a science, a farming method, and a movement. As a science and method, agroecology seeks to integrate ecological concepts and principles into the design and management of sustainable farming systems (Altieri 2019). Agroecologists promote the diversification of farming systems to minimize dependence on external inputs and increase productivity and resilience. For example, polycultures, crop-livestock combinations, rotations, and agroforestry systems, among other management practices. For rural social movements in Latin America and beyond—as advanced by the transnational peasant movement, La Vía Campesina—agroecology is more than the technical dimensions of the farming method: it is a radical movement for social, environmental, and climate justice (Altieri and Toledo 2011; Martínez-Torres and Rosset 2014; Svampa and Viale 2020).

Agroecology as a scientific discipline had its origins between the 1930s and 1960s in Russia and Germany (Wezel et al. 2009). Agroecology as a science and a practice reemerged in Latin America in the 1980s (Altieri and Nicholls 2017). The field, traditionally dominated by agronomy and ecology, was influenced by critical social sciences, such as anthropology, rural sociology, and development studies, which called to center the experiential agroecological knowledge of farmers. During the 1980s and 1990s, agroecology expanded across the region owing to nongovernmental organizations (NGOs) adopting agroecological methods to address the negative social and ecological impacts of the Green Revolution. The NGOs saw agroecology as an alternative to inefficient top-down approaches to address rural development, poverty, and land degradation. Since the turn of the twenty-first century, agroecology has expanded rapidly throughout Latin America thanks to new forms of collaborations between rural movements, NGOs, and academics that blend agroecological science and Indigenous knowledge systems (Altieri and Nicholls 2017; Altieri and Toledo 2011).

Agroecology as a movement emerges as a challenge to the expansion of the agribusiness frontier into Indigenous and peasant territories (Rosset and Martínez-Torres 2012). As Diana Stuart argues in chapter 14, while the global environmental crisis can feel overwhelming, falling into defeatism only serves to maintain the status quo. Movements like La Vía Campesina (LVC) demonstrate how local initiatives can effectively challenge corporate power while building real alternatives. LVC is an international movement composed of social movements and organizations of peasant and family farmers, Indigenous peoples, rural women and youth, landless peasants, and farm workers worldwide. By 2022, its thirtieth anniversary, LVC comprised 182 local and national organizations from eighty-one countries across Asia, Africa, Europe, and the Americas (La Vía Campesina 2023). LVC has been strategic in promoting agroecology as a farming method to recover soils degraded by industrial agriculture and regain peasant autonomy by minimizing dependence on external inputs. As Rosset and Martínez-Torres (2012, 2014) argue, LVC has been particularly strategic in promoting agroecology as a framing for collective action for Indigenous and peasant control over their territory and their ways of knowing and being in the world.

Agroecology as a movement thus encompasses concrete struggles over the natural resources that sustain life and a challenge to the "monocultures of knowledge." Compared to top-down, profit-motivated, expert approaches to farming imposed by the agribusiness model, agroecology as a movement advances a horizontal approach to sharing knowledge built on previously undervalued local and Indigenous knowledges and worldviews. Agroecology as a movement expands and is sustained through *diálogo de saberes* and *campesino-a-campesino* networks and processes. *Diálogo de saberes* is a concept and a method implemented by LVC member movements in Latin America that translates as "dialog among different knowledges and ways of knowing" (Martínez-Torres and Rosset 2014, 980). As Enrique Leff (2011) argues, *diálogo de saberes* is central to environmental epistemology (that is, how we know and understand the environment), as it emphasizes the importance of bringing together different forms of knowledge, including scientific, traditional, Indigenous, and local knowledge, to address environmental challenges. It invites creating a space for dialogue and collaboration between diverse knowledge systems, recognizing the need to move beyond a purely scientific approach to environmental issues and incorporate the wisdom and knowledge embedded in different cultural, social, and ecological contexts. That is, as Michael Warren Murphy (chapter 7) argues, to emphasize insights gathered from subaltern standpoints. The *campesino-a-campesino*, or farmer-to-farmer, movement, is a peasant-driven process of technological innovation that originated in the highlands of Guatemala in the 1980s from Kaqchikel Mayans and Mexican farmers (Holt-Giménez 2006). This movement recognizes the expertise that farmers have gained through their own experiences and emphasizes the importance of a horizontal exchange of knowledge and skills. Since the 1980s, the *campesino-a-campesino* methodology has been strategic to disseminating and scaling up agroecological practices across Central America, Mexico, Cuba, and Brazil (Altieri and Toledo 2011; Bernal et al. 2023).

Diálogo de saberes, as a methodology, has served to promote agroecology as a farming method and for the collective construction of *food sovereignty* as a collective frame for resistance (Martínez-Torres and Rosset 2014). La Vía Campesina defined **food sovereignty** in the 2007 *Declaration of Nyéléni* as:

> the right of peoples to healthy and culturally appropriate food produced through ecologically sound and sustainable methods, and their right to define their own food and agriculture systems. It puts the aspirations and needs of those who produce, distribute and consume food at the heart of food systems and policies rather than the demands of markets and corporations. It defends the interests and inclusion of the next generation. It offers a strategy to resist and dismantle the current corporate trade and food regime, and directions for food, farming, pastoral and fisheries systems determined by local producers and users. Food sovereignty prioritises local and national economies and markets and empowers peasant and family farmer-driven agriculture, artisanal—fishing, pastoralist-led grazing, and food production, distribution and consumption based on environmental, social and economic sustainability. Food sovereignty promotes transparent trade that guarantees just incomes to all peoples as well as the rights of consumers to control their food and nutrition. It ensures that the rights to use and

> manage lands, territories, waters, seeds, livestock and biodiversity are in the hands of those of us who produce food. Food sovereignty implies new social relations free of oppression and inequality between men and women, peoples, racial groups, social and economic classes and generations. (La Vía Campesina 2007)

Food sovereignty stands in opposition to the calls for food security advanced by corporate leaders, government officials, and international organizations like the United Nations. Food security refers to "access to sufficient, safe, and nutritious food," while food sovereignty explicitly targets the right to social control over the food system (Patel 2009). Food sovereignty challenges unequal power dynamics over the control of natural resources and the narratives that sustain those dynamics. In opposition to "food from nowhere," the call for food sovereignty shortens the distance between producers and consumers by relocating food production processes.

CONCLUSION

The future of food is at a crossroads, navigating divergent paths shaped by competing ideologies and approaches. The historical trajectory from the colonial origins of export monocultures to today's corporate-driven agriculture reveals the evolution of our food provisioning systems and the deep-seated inequalities and ecological ramifications resulting from the constant expansion of agrarian frontiers for capitalist accumulation.

The expansion of these agricultural frontiers, marked by the monoculture mentality of Western modernity, has given rise to a corporate food regime that prioritizes efficiency, profitability, and technological innovation. From robot weeders in Arkansas to lab-grown meat in California and the knowledge-based network model in Argentina, the narrative of ecological modernization claims to deliver a future where science and technology alleviate the environmental impacts of agro-industrial practices.

However, this path raises critical questions about unequal power dynamics in decision-making and the distribution of costs and benefits. The increasing digitalization and financialization of the food system, coupled with a growing disconnect between producers and consumers, underscore the need for a more just approach. As we consider the implications of a future reliant on labs, robots, and venture capital, it becomes apparent that efficiency alone cannot address the underlying issues of social justice, environmental sustainability, and global inequality embedded in our food system.

In contrast, the agroecological movement emerges as a beacon of hope and transformative potential. From its roots as a science and method in the twentieth century, agroecology has evolved into a dynamic movement championed by Indigenous and peasant movements, challenging the relentless expansion of the agribusiness frontier and advocating for social, environmental, and climate justice. The integration of ecological principles, the emphasis on local and Indigenous knowledges, and the promotion of food sovereignty redefine the contours of a just and sustainable food future.

The agroecological movement, propelled by initiatives such as *diálogo de saberes* and *campesino-a-campesino* networks, not only seeks to recover degraded soils and promote resilient farming systems but also challenges the prevailing "monocultures of knowledge." The call for food sovereignty articulated by La Vía Campesina stands in stark contrast to the narrow focus on food security, advocating for the right to social control over the entire food system.

The chapters in this third and final section of this volume reveal how contemporary power structures, technological systems, and social inequalities are deeply intertwined in shaping our ecological relationships and possible futures. At stake is not just *who* controls institutions and technologies but *how* these shape our relationships with the environment and with each other. The divergent paths for the future of food exemplify these interconnections: ecological modernization represents the continued dominance of corporate institutions and their approach to technological innovation, while the agroecological movement offers an alternative that fundamentally challenges existing power structures while advancing different forms of knowledge and technology in service of social and ecological justice.

As we stand at the crossroads of these two divergent paths, the imperative is clear. The future of food necessitates a paradigm shift—one that moves beyond mere efficiency to embrace justice, sustainability, and inclusivity. By acknowledging the voices of those often marginalized in the current food discourse, fostering collaboration between different knowledge systems, and championing the principles of agroecology and food sovereignty, we can collectively forge a path toward a more resilient, more just, and more sustainable global food system.

NOTES

This chapter emerged from discussions with students in my Sociology of Food and Agriculture course at Tulane University, developed with support from the Rosenthal Blumenfeld Gulf South Food Studies Fellowship.

1. Harriet Friedmann and Philip McMichael (1989) developed Food Regime (FR) analysis as a theoretical framework to explain the rise and decline of national agricultures within the geopolitical history of capitalism. The framework emphasizes

the historical dimensions of how global food provisioning emerged and has changed substantially across time and space. The First FR was British-centered (1870s–1930s), combining colonial tropical imports to Europe with basic grains and livestock imports from settler colonies. The Second FR was US-centered (1950s–70). The United States gained hegemony via subsidizing key crops at home and rerouting surplus as food aid and exporting the agricultural technologies of the Green Revolution to strategic states on the Cold War perimeter. The Third Corporate FR (1980s onward) has been dominated by transnational corporations that have established a regime based on the commodification of food provisioning. For more details, see McMichael (2009, 2013).

2. I take from the companies' websites to describe the process of lab-grown meat. See "How We Make Meat," GOOD Meat, https://www.goodmeat.co/process; "Cultivated Meat. It's Science (but Not Rocket Science)," UPSIDE Foods, https://upsidefoods.com/innovation.

3. "We're Making Our Favorite Food a Force for Good," UPSIDE Foods, https://upsidefoods.com/company.

REFERENCES

Altieri, Miguel A. 2019. "Agroecology: Principles and Practices for Diverse, Resilient, and Productive Farming Systems." *Oxford Research Encyclopedia of Environmental Science*. Oxford University Press.

Altieri, Miguel A., and Clara I. Nicholls. 2017. "Agroecology: A Brief Account of its Origins and Currents of Thought in Latin America." *Agroecology and Sustainable Food Systems* 41 (3–4): 231–37.

Altieri, Miguel A., and Victor Manuel Toledo. 2011. "The Agroecological Revolution in Latin America: Rescuing Nature, Ensuring Food Sovereignty and Empowering Peasants." *Journal of Peasant Studies* 38 (3): 587–612.

Bernal, David, Omar Felipe Giraldo, Peter M. Rosset, Oliver Lopez-Corona, and Julian Perez-Cassarino. 2023. "Campesino a Campesino (Peasant to Peasant) Processes versus Conventional Extension: A Comparative Model to Examine Agroecological Scaling." *Agroecology and Sustainable Food Systems* 47 (4): 520–47.

Buttel, Frederick H. 2000. "Ecological Modernization as Social Theory." *Geoforum* 31 (1): 57–65.

Clapp, Jennifer. 2014. "Financialization, Distance and Global Food Politics." *Journal of Peasant Studies* 41 (5): 797–814.

Fairbairn, Madeleine. 2014. "'Like Gold with Yield': Evolving Intersections between Farmland and Finance." *Journal of Peasant Studies* 41 (5): 777–95.

Friedmann, Harriet, and Philip McMichael. 1989. "Agriculture and the State System: The Rise and Decline of National Agricultures, 1870 to the Present." *Sociologia Ruralis* 29 (2): 93–117.

Ghosh, Amitav. 2022. *The Nutmeg's Curse: Parables for a Planet in Crisis*. University of Chicago Press.

Harrison, Jill Lindsey. 2011. *Pesticide Drift and the Pursuit of Environmental Justice*. MIT Press.

Holt-Giménez, Eric. 2006. *Campesino a Campesino: Voices from Latin America's Farmer to Farmer Movement for Sustainable Agriculture*. Food First Books.

Lander, Edgardo, ed. 2000. *La Colonialidad del Saber: Eurocentrismo y Ciencias Sociales. Perspectivas Latinoamericanas*. CLACSO, Consejo Latinoamericano de Ciencias Sociales.

La Vía Campesina. 2007. "Declaration of Nyéléni." Nyéléni. https://nyeleni.org/en/declaration-of-nyeleni/.

La Vía Campesina. 2023. *La Vía Campesina: 2022 Annual Report*.

Leff, Enrique. 2011. *Aventuras de la epistemología ambiental: De la articulación de ciencias al diálogo de saberes*. Siglo XXI.

Leguizamón, Amalia. 2014. "Modifying Argentina: GM Soy and Socio-Environmental Change." *Geoforum* 53:149–60.

Leguizamón, Amalia. 2016. "Disappearing Nature? Agribusiness, Biotechnology and Distance in Argentine Soybean Production." *Journal of Peasant Studies* 43 (2): 313–30.

Leguizamón, Amalia. 2020. *Seeds of Power: Environmental Injustice and Genetically Modified Soybeans in Argentina*. Duke University Press.

Little, Amanda. 2019. *The Fate of Food: What We'll Eat in a Bigger, Hotter, Smarter World*. Harmony.

Lynch, John, Michelle Cain, David Frame, and Raymond Pierrehumbert. 2021. "Agriculture's Contribution to Climate Change and Role in Mitigation is Distinct from Predominantly Fossil CO2-Emitting Sectors." *Frontiers in Sustainable Food Systems* 4.

Martínez-Torres, María Elena, and Peter M. Rosset. 2014. "Diálogo de Saberes in La Vía Campesina: Food Sovereignty and Agroecology." *Journal of Peasant Studies* 41 (6): 979–97.

McMichael, Philip. 2009. "A Food Regime Genealogy." *Journal of Peasant Studies* 36 (1): 139–69.

McMichael, Philip. 2013. *Food Regimes and Agrarian Questions*. Fernwood.

McMichael, Philip, and Heloise Weber. 2021. *Development and Social Change: A Global Perspective*, 7th ed. Sage.

Mintz, Sidney Wilfred. 1986. *Sweetness and Power: The Place of Sugar in Modern History*. Penguin.

NPR 1A. 2023. "What Is Cultivated Meat?" NPR. https://www.npr.org/2023/09/05/1197653592/what-is-cultivated-meat.

Oliveira, Gustavo, and Susanna Hecht. 2016. "Sacred Groves, Sacrifice Zones and Soy Production: Globalization, Intensification and Neo-Nature in South America." *Journal of Peasant Studies* 43 (2): 251–85.

Patel, Raj. 2009. "Food Sovereignty." *Journal of Peasant Studies* 36 (3): 663–706.

Patel, Raj. 2013. "The Long Green Revolution." *Journal of Peasant Studies* 40 (1): 1–63.

Patel, Raj, and Jason W. Moore. 2017. *A History of the World in Seven Cheap Things: A Guide to Capitalism, Nature, and the Future of the Planet*. University of California Press.

Rosset, Peter, and Maria Elena Martínez-Torres. 2012. "Rural Social Movements and Agroecology: Context, Theory, and Process." *Ecology and Society* 17 (3).
Shiva, Vandana. 1993. *Monocultures of the Mind: Perspectives on Biodiversity and Biotechnology*. Illustrated edition. Zed Books.
Soluri, John. 2005. *Banana Cultures: Agriculture, Consumption, and Environmental Change in Honduras and the United States*. University of Texas Press.
de Sousa Santos, Boaventura. 2014. *Epistemologies of the South: Justice Against Epistemicide*. Routledge.
Svampa, Maristella, and Enrique Viale. 2020. *El colapso ecológico ya llegó: Una brújula para salir del (mal)desarrollo*. Siglo XXI.
UN News. 2023. "Hunger Afflicts One in Ten Globally, UN Report Finds." July 12.
Wezel, Alexander, S. Bellon, T. Doré, C. Francis, D. Vallod, and C. David. 2009. "Agroecology as a Science, a Movement and a Practice. A Review." *Agronomy for Sustainable Development* 29 (4): 503–15.

13. Imperial Cities and Climate Change

Hillary Angelo

While urban and environmental sociology have traditionally ignored each other, climate change's threats to urban assets and cities' centrality as sites for climate solutions are prompting urbanists to engage environmental questions, and vice versa. But how should the environment figure into contemporary urban analysis? What relationship do urbanists and urbanism have to environmental sociology, and socio-environmental problems, today? These aren't just intellectual problems but are related to real questions that climate change raises. How should we think about the urban ecology of the twenty-first century? What new material, political, and energetic relations between city and countryside might be built as we decarbonize? And how can planners and policymakers best understand and address these questions?

This chapter treats current crises as an opportunity for reflection on one historical point of intersection between urbanization and the environment, and between urban and environmental analysis: the "imperialism" of cities and metropoles over regional and global hinterlands (Brechin 2006). I argue that while certain features of contemporary socio-environmental problems are new, this basic, persistent material and ideological relation is not—and that it is essential to understand this dynamic as we work to rebuild energy, water, and food systems in response to climate change and in the pursuit of just, sustainable urban futures.

The chapter tells a brief history of urban imperialism through the growth of Western cities in North America, highlighting the political and material asymmetries that have characterized capitalist urbanization over the past two centuries, and with a focus on the recurrent marshaling of extra-urban resources for urban life. First, urban needs: provisioning for city populations has long motivated wide-ranging transformations in land use and natural resource management. Second, urban power: capital and interests of those located in cities have long had a hand on the lever of decision-making, extracting from the countryside and accumulating wealth, waste, and political power. And third, urban imaginaries: ideals of modern, urban futures influence the conceptualization of problems and solutions in popular and technocratic realms.

I then reflect on these patterns in the context of climate change and current scholarship. In requiring coordinated land use change across country and city, and given the injustices and unsustainabilities of this pattern, climate change prompts a reconsideration of these dynamics and is an opportunity to transform them. This is, therefore, a moment to reflect on past relations as we are making decisions about the future, and a moment in which it is particularly important to bring this historical picture into focus.

ENVIRONMENTALIZING CITIES?

Cities are environmental beasts. As is now commonly quipped, skyscrapers are no less "natural" than a tree (Harvey 1996). The material environment of cities' roads, buildings, and infrastructure is earth and metals transformed; meanwhile, urban life is reliant on ecological resources brought in from outside—food, water, and labor—and needs sinks for waste, pollution, and other excesses beyond. Yet despite this ongoing metabolic transformation and exchange, urban life can and often has appeared to be an island. And urban life of the past couple of centuries has been marked by a particular "amnesia" when it comes to industrial cities' reliance on their hinterlands (Castree 2000).

There are good reasons that modern urbanites continually forget about the environment on which cities depend. In Western Europe and the United States, industrial metropolises were described as accomplishments of human "domination over nature." Politically and epistemologically, nature was understood to lie "outside" them. Classical sociology took nature as the site of society in contrast to nature, the rural, the "not-yet" city; urban studies (and arguably sociology itself) enshrined modernity's dualisms—city/nature, society/nature, *Gemeinschaft/Gesellschaft*—in its disciplinary formation and objects of analysis (see Angelo 2017). And bifurcations between urban and environmental studies solidified these disparate understandings in contemporary scholarship as well (Angelo and Greenberg 2023). For urbanists and sociologists, cities became the site of society; for environmentalists, they became environmental problems to be "solved" (Angelo and Wachsmuth 2020).

These associations have in many ways been destabilized by climate change. Sea level rise, heat, and storms and hurricanes mark the apparent "return" of wild nature to the city in popular discourse. There's also been a remarkable "environmentalization" of urban policy over the past fifty years, as cities have become key sites for the pursuit of sustainability (Angelo and Wachsmuth 2020). And, with climate change, come new urban environmental challenges, among them the urgent need to rebuild the systems that bring water, food, and energy into cities in ways that reduce emissions and lower energy use.

Against the background of modern, Western understandings of cities as opposed to nature, all this attention to the environmental dimensions of urban life seems very new. The environmentalization of cities is sometimes described in media and popular culture as a break from recent history, given the urban discourses of the recent past. And yet the basic questions climate change raises—about cities' relationship to their hinterlands, to wild nature, and to ecological resources that lie outside—are very old. Despite the newness of specifically climate-related problems and questions, basic contours of the "environment's" material and ideological entanglements with cities and urbanism are not changed much by these contemporary dynamics.

One scholarly response to perceptions of the environment as "outside" cities in media understanding and popular discourse has been efforts by urban historians and political ecologists to "show" the nature of cities and analyze interdependencies between city and hinterland, colony and metropole, and flows of goods, people, and energy between (e.g., Heynen et al. 2006; Gandy 2003). These have been important efforts, in recent decades, to remind scholars of the environmental content of city streets and skyscrapers, and to critique ideological (and frequently anti-urban) notions of cities as unnatural and antithetical to environmental concerns or ways of living.

However, in the course of deconstructing these ideologies, another environmental dynamic has been backgrounded. A city is no less "natural" than a tree, but in creating these environments a recurrent pattern of infrastructure development, resource use, and urban form has been as follows: cities pull resources into their orbits to build up infrastructure, stabilize buildings and living space, and secure resources to sustain life. In the course of enriching and expanding urban life over the past 150 years, rural lands and lives have been impoverished and marginalized. This uneven ecological relationship is built in to the physical shape of infrastructure and urban form, and it results in deeply bifurcated experience (and understanding) across country and city.

We should return to this historical dynamic now, in the context of climate change, as the decisions it prompts, and new infrastructure and land uses it requires, have the chance to further cement these political and metabolic relations between cities and countryside or, potentially, to transform them.

THE AMERICAN WEST AS A CASE OF URBAN IMPERIALISM

It is well known today that, internationally, colonialism and imperialism created global "hinterlands" that served as extractive resource frontiers and that have continued to be sites of low wages, cheap manufacturing, and polluting industries, mostly in the so-called Global South, for the accumulation of capital in the Global North (Grove 1996; Arboleda 2021; see the introduction and chapter 7 of this volume). The historian Gray Brechin's (2006) concept of *urban* imperialism highlights the fact that during the same centuries, cities have shared similar relational dynamics with their domestic, adjacent hinterlands.

For at least two centuries, capitalist urbanization has featured a distinct metabolic and political pattern: the subordination and instrumentalization of the countryside to cities and their needs. While not always articulated in such terms, a variety of concepts commonly used to describe unequal environmental relationships between city and countryside—for example, the metabolic rift (Foster 1999) or extended urbanization and "operational landscapes" (Brenner and Schmid 2015), as well as "imperial" cities (Brechin 2006)—all contain elements of such an understanding. This relationship is a result of urban needs, urban power, and urban imaginaries marshalling extra-urban resources for city life.

Urban Needs

The most basic of these connections between cities and the "environment" have been the food, water, and other material resources (energy! labor!) brought in to cities to make urban life go. The environmental historian William Cronon's *Nature's Metropolis* (1991) was a landmark account of Chicago's reliance on the natural resources of its hinterland for urban growth. As he and other urban-environmental scholars have since emphasized, provisioning for city populations has required large-scale, hugely expensive, and deeply political transformations in land use and natural resource management. Rather than a simple harvest of natural assets that offered themselves up for urban consumption, there was a lot of human work involved in instrumentalizing nature for urban needs. Rivers had to be dredged; ports had to be created; trees and ore had to be transformed into building materials.

The classic example of this is water. Large settlements where people were not engaged in subsistence agriculture preceded industrial cities, and these required water to drink, to remove waste, and for agriculture. Some water infrastructures are relatively direct extensions of "found" natural systems, such as water wheels or irrigation channels, but as technological capacities have grown, water flows and systems have become more heavily modified. More extensive and technologically complex water infrastructure has meant that the size and location of human settlements are less constrained by "natural" patterns. Because we can transport drinking water and liquid waste for miles, we can sustain populations that are too large for proximate sources. Because we can pump water over mountain ranges, we can build cities in arid deserts.

The improbable cities of the American West are emblematic of this relationship. Los Angeles, in the words of John Logan and Harvey Molotch (2007), "had none of the 'natural' features that are thought to support urban growth: no centrality, no harbor, no transportation crossroads, not even a water supply. Indeed, the rise of Los Angeles as the preeminent city of the West, eclipsing its rivals San Diego and San Francisco, can only be explained as a remarkable victory of human cunning over the so-called limits of nature" (55). Los Angeles—like, we might add, Phoenix, Denver, and Las Vegas—is a "beautiful fraud" (Reisner 1993, 344), an absurd oasis that owes its existence to "the most elaborate hydraulic system in world history, overshadowing even the grandiose works of the Sassanians and the Pharaohs" (Worster 1982, 56).

Los Angeles was able to grow and thrive only because of extensive water projects in California's Central Valley that brought drinking water to the city. In the 1870s and 1880s, there was still relatively easy water access, and enough private dollars—from individuals or private companies—to secure the water that was needed. By the mid-1920s, there was no more easy water. The Bureau of Reclamation was founded in 1902 by the Secretary of the Interior (and funded by revenue from the sale of federal lands) for the purpose of water development in Western states. The infrastructure and irrigation projects it funded transformed California's arid Central Valley into the largest expanse of irrigated farmland in the world (Reisner 1993, 348), and produced what had been, until recently, the two biggest irrigation projects on earth: the federally funded Central Valley Project and the state-funded California State Water Project.

The Central Valley Project began at end of the 1930s, when the ground water table was dropping precipitously after almost a decade of drought.

The need for alternatives to groundwater was dire enough that, even in the middle of the Great Depression, pressure from farmers convinced the federal legislature to authorize an eighteen-year project to construct series of reservoirs to store water in the wet northern half of the state and transport it to the dry Central Valley by canal (Worster 1992, 10). In the 1950s, the California State Water Project increased irrigation in the Central Valley and brought more drinking water to Southern California cities. Like the Central Valley Project, the California State Water Project brought more Northern California water south, but was far more elaborate. It pumped water hundreds of miles via aqueducts to irrigate arid areas, and over a mountain range to get water to Southern California cities. In the 1980s, inflation hugely escalated the costs for its completion (373–74), making it the largest state-financed water project ever built (Water Education Foundation 2023).

Hydraulic control over arid lands made California's Central Valley into "the world's fruit basket," Los Angeles into a city of 3.5 million, and the state's into the sixth largest economy in the world. Today, as California faces historic droughts and intensifying wildfires, the state continues to produce more than half the fruit, vegetables, and nuts consumed in the United States each year. These come from 750,000 acres of desert farmland in the Central Valley that have been "reclaimed" through water infrastructure. Almonds and oranges are important but, as Los Angeles voters foresaw, residents of Southern California cities literally could not survive without these water projects today; 70 percent of the California State Water Project's flow supplies drinking water to Southern California cities (Water Education Foundation 2023). And one axis of contemporary battles over water allocation in California—as across the Southwestern United States generally—is whether urban (drinking water, golf courses, pools) or agricultural uses (which feed cities indirectly and bring income to farmers) should be prioritized.

Urban Power

Decisions about the manipulation and use of natural resources such as water are deeply political. They involve not just technological but political power, as well as choices about who sacrifices, who should pay what price, and who decides all along the line. Those decisions that have accompanied the transformations of the industrial era have repeatedly resulted in extraction from, wasting of, and impoverishment of the countryside to support urban growth.

In a classic story, the modern transition to a market society, and, later, an urbanized, industrial capitalist society, can be understood as the "subordination of land to the needs of a swiftly expanding urban population" (Polanyi 2001, 189; see also Thompson 1963), which was accomplished by the exercise of public and private urban power. While global colonial violence extracted natural resources and coerced labor, in England, the state participated in "enclosures," or the privatization of common land, which pushed peasants off the land and into cities for employment. Enclosures ended open grazing and pasture, as well as smallholdings, and resulted in the consolidation of land ownership. As landless peasants moved to cities where factory owners waited with jobs in hand, private landowners engaged in "improvements" to make the larger parcels more productive, in order to feed the growing industrial workforce and urban elites. This included draining fens and marshes to create more arable land, and adding fertilizers and other mechanisms to improve productivity. European colonial powers repeated these patterns abroad in plantations that imported goods like sugar and tobacco for urban consumption. All these changes consolidated the "industrial-agricultural division of labor" by the nineteenth century (Polanyi 2001, 190) and created antagonistic country-city dynamics domestically and internationally, while impoverishing rural peasants and the new urban proletariat. In the countryside, the state responded with poor laws, a kind of welfare for those suffering, while the rise of urban poverty led to charity and social reform movements—with both states and private citizens asking themselves where all these poor people had come from. While we might not usually think of such control and channeling of labor as analogous to the control and channeling of natural resources, these efforts all made both land and waged and unwaged labor expropriatable, instrumentalized them for the "best" human use, and modified and dislocated them in the service (primarily) of accumulation, and (secondarily) of urban consumption (see Moore 2015; Federici 2004).

Fast forward to the twentieth century, and cities and states continue to rely on an ecological basis of social power. Increasing technological and administrative capacities have moved us from such crude efforts to control land and natural resources to increasingly complex ones, but the basic rule applies: ecological power is social power. German historian Karl Wittfogel (1957) once described this dynamic, in the context of water, as a "hydraulic state," in which technological control of water translates into political power. Wittfogel was concerned with a specific "hydraulic order of life" in arid landscapes (12), particularly irrigation systems in Mesopotamia and Egypt. As the scale of water management escalated, he argued, political power come to rest in the hands of a bureaucratic elite who neither owned the land or water but became the ruling class through control of infrastructure that mediated access to it. Citizens had no choice but to go through the state's physical infrastructure, political institutions, and bureaucratic structures to meet their needs.

This basic insight applies to contemporary arid lands—such as the "water wars" in the American Southwest described above—and to city-hinterland dynamics. Once again, there's been a recurrent spatial dimension to these politics of power in the modern era, in which the resources of the countryside are marshaled for specifically urban use and consumption. The hydraulic state works in tandem with local boosters, urban elites, and city growth machines to invest in, and funnel resources toward, cities. Wittfogel did not describe how industrial capitalists, a modern class of urban elites, might have pressured states to make certain kinds of decisions. But what Logan and Molotch (2007) famously termed urban "growth machines" have been key sites of urban power: forces that pull profit, people, and other natural resources to them, while throwing off waste outside.

The funding for the California water projects described above was public state and federal dollars; these projects were designed by local real estate and business elites who were able to secure federal funds to bring water and power to the city. Beyond drinking water, federal funding was also secured to dredge the Los Angeles port, which was a basic condition for making the city viable as a large metropolis. Both these "infrastructural victories" were won by city boosters, who managed to secure millions in federal funds for infrastructure projects (Logan and Molotch 2007). For its authorization, the California State Water Project also required the support of prescient Los Angeles voters who anticipated future water needs as the city continued to grow (Worster 1992, 355). And, as the classic film *Chinatown* shows, resource and power struggles didn't end once the infrastructure was completed. The Southern California "water wars" began as the city outgrew its water supply in the 1930s and had to start diverting water from the Owens Valley, forcing farmers to sell their land or divert water away from farming and ranching. A variety of sketchy land deals were used to continue to transfer land and water rights to the city of Los Angeles throughout the twentieth century (Reisner 1993; Kahrl 1982).

In these ways, urban power has far-reaching environmental consequences. Brechin's *Imperial San Francisco* (2006) describes the city as an inverted mine—shafts into skyscrapers—through the "conversion of nature's wealth into city wealth" (20, 32, 17). Beyond mines, rivers become drinking water, or else are channeled into almonds or alfalfa. Alfalfa feeds cattle that become hamburgers, or steaks in high-end restaurants. Coal and oil power homes, or factories that make iPhones. Lithium mined in remote areas becomes batteries powering electric cars. Thus, city-hinterland relations are very much a domestic analogue of metropole-colony relationships. The marshalling of nearby water to grow Los Angeles, or mining to enrich San Francisco, or forcing peasants into cities to labor in factories is very much like the extraction of natural resources and enslavement in colonies, right down to the environmental degradation that has resulted from such schemes (Holleman 2018).

Brechin describes the impacts of city-building on regional and international hinterlands as "ecocide" and takes pains to put faces on the urban capital and interests who make decisions: flash-in-the-pan speculators, but also the "dynastic elites" of the urban aristocracy, such as, in San Francisco, the Scott brothers, the De Youngs, and the Hearsts, for whom rural extraction is a means to consolidate urban wealth and power. He argues that in these ways, contemporary imperial cities—much like ancient city-states—are still pulling resources into their orbits, and decisions are still being made about hinterlands from the vantage points (and reflecting the priorities of) these urban centers. These dynamics also reflect the fact that power is concentrated in cities, and so the wresting of nature is recurrently in one direction: flowing from country to city, hinterland to metropole, mine to skyscraper—rural extraction powering urban development.

Urban Imaginaries

A third element of urban imperialism comes in the form of urban imaginaries, which influence the conceptualization of problems and solutions in the popular and technocratic realms. As described above, decisions about land use, labor, and the direction and extension of natural resources of this era have had a particular political geography. For the most part, each brought resources from country to city. Each contributed to the emptying out of rural areas and, in that vacuum, to the projection of urban imaginaries onto rural or agrarian spaces. The organization of the systems—of labor (urban wage labor and consumption; production abroad), of infrastructure (large scale, long-distance) and of the design of urban space (e.g., the hiding of slaughter on city margins)—that supports and has enabled these settlements has also created distinctly different experiences of these systems and processes across country and city. This, in turn, has contributed to a bifurcated understanding of these issues across urban and rural space.

Urban environments create urban imaginaries (Angelo 2021). The infrastructure that helps meet city populations' needs for food, energy, and water doesn't only marshal resources but plays a key role in perception, including of views of nature, the countryside, and the environment "outside." In the case of water, the same systems that make water available at the spin of the tap also—in obscuring rural sources—make urbanites far less conscious of where that water might have come from than, for instance, someone who must fetch water daily from a river or well (Kaika 2004). Such highly mediated ecological relations permeate urban life, in forms such as prepared food from a grocery store or fast travel behind the glass windows of a car or train instead of by horse or on foot. In Brechin's (2000) words, "the paradox of the imperial city is that as it grows, its inhabitants become ever less aware of their own growing dependence and impact on the city's expanding contado" (516; "contado" is the Italian word he uses for a city's hinterland or periphery). Large-scale infrastructure of this kind reinforced the idea of cities' separation from nature by decreasing urbanites' direct access to and awareness of the extended networks and resources on which urban life depends.

In addition to making it hard to discern the resources and systems people are reliant on, these environments and systems also distort urbanites' views of the hinterland—and rural residents' views of the city. For city people, they tend to result in romanticized pastoral ideals of rural idylls and, at the same time, schemes to urbanize, industrialize, and modernize the countryside.

Like urban needs and power, these urban imaginaries have a long history. Raymond Williams's *The Country and the City* is perhaps the most famous portrait of the ever-present nostalgia for disappearing rural life. Through poetry and prose, Williams tracks recurring romantic sentiments of countryside as a utopian, Edenic, anti-modern Golden Age, and shows the roots of urban transformations continually understood to have sullied rural life *in the countryside*, pointing out that the socio-ecological transformations of city and countryside have gone hand in hand.

Similarly, imaginaries of natural idylls are constructed *through* urban life; material relations helped produce imaginative ones. As cities became the sites of "society," of modernity, of work, of social troubles, the rural or the non-city became a playground for urbanites: once people worked in factories rather than on farms, they might seek out weekends in the country for pastoral forms of leisure. Rural nature as recreation comes into being in industrial, urban environments: farmlands and fields can look relaxing when work takes place indoors; wild nature can look fun once real threats (of freezing, starvation, survival, etc.) have been removed. These experiences are, of course, also mediated through the various stratifications of urban life—particularly, race, ethnicity, and class. Nature consumption requires sufficient discretionary income to purchase cars and access these places, as well as a cultural and social history of "belonging" in such spaces (Taylor 2016; Finney 2014). Affluence, security, and related pastoral pursuits (hunting, golfing, running) have disproportionately been the property of white, middle- and upper-class urbanites, resulting in contemporary efforts to "democratize" nature experience—to make these places more physically and socially accessible to a wider range of people.

At the same time, and existing in parallel with this desire to consume and romanticize urban space, has been the impulse to modernize and "improve" it. In the nineteenth century and throughout the twentieth, with cities understood to be the site of society, the rural/agrarian world became the "not-yet" city—a land in need of development (Robinson 2002). Efforts to "modernize" rural life have included the direct provision of goods, services, infrastructure, and education, as well as philanthropic investments hoping to encourage employment, industrialization, entrepreneurship, and so on. But the forms of rural underdevelopment that such efforts target are, in fact, *features* of particular forms of capitalist urban development. The historian Andrew Needham (2014) has described how "metropolitan development" was supported by "Navajo underdevelopment" in the Four Corners region of the United States. In addition to water that was taken from the Colorado River (and other rivers), slowly parching nearby communities, power stations sited on reservation land sent ample and cheap power to cities, while the tribal communities adjacent to the original bodies of water were unserved, or underserved, by the same facilities—and then directly bore the costs. And, as in the case of international development, such exploitative practices have coexisted with targeted, simultaneous schemes to educate, assimilate, and modernize Indigenous populations. At best, such programs have been well intentioned, if not always effective. At worst, they have been forms of predatory inclusion that peddle undesirable land uses in rural areas, from prisons to energy infrastructure, with dubious local benefits.

Of course, this trajectory has been far more complex, varied, and variable than I have depicted it here. Such transformations have been resisted by rural populations and critiqued by social and environmental scholars. And yet, in the main, over the past 150 years we've seen an emptying out and impoverishment of the countryside and consolidation of resources, power, and people in cities. Today, we live in a world where cities are sites of power and resources, where most people go for lives and livelihoods, where the majority of the world's poverty is rural, and where urban needs, power, and imaginaries have directed resource use and shaped socio-ecological environments within and beyond cities. Urban imperialism even extends to disciplinary configurations as well—it's apparent in the separation between urban and rural sociology, and in rural (and more broadly environmental) sociology's continued marginalization by the rest of the discipline.

DISRUPTING THESE DYNAMICS? DEMOCRACY, SUSTAINABILITY, AND RELATIONALITY IN A TIME OF CLIMATE CHANGE

Enter climate change. The ecological questions prompted by climate change have played a part in destabilizing the "amnesia" of cities' dependence on nature in the form of hinterlands, labor, water, and so on, rather than actually being any kind of new "environmentalization" of cities themselves. Yet, insofar as a large part of climate adaptation requires coordinated land use decisions across country and city, climate change nevertheless raises a number of environmental questions that have been settled for a time about energetic relations between urban and rural land and life. As the ecological conditions within which city-hinterland relations are embedded change, questions such as the following arise: What to do with cities and/or real estate markets, which "followed" infrastructure development that gathered people and commerce in specific places, in a climate-changed future? What if drought and shrinking allocations from the Colorado River mean that drinking water can no longer be guaranteed? Should Los Angeles simply be abandoned? Should California agriculture be sacrificed? Or should the river's remaining water be left to replenish parched wetlands and tributaries, shrinking underwater aquifers, and dying desert hot springs? Or, should huge new water projects be built—moving water from the Mississippi River to Lake Powell and Lake Mead, or from the Pacific Ocean to refill the Great Salt Lake? Which of these uses and environments should be prioritized? What are the costs of such decisions? Who pays the price? And who decides?

These questions reprise the old choices and dynamics of the past two centuries. Hinterlands' subservient relation to cities, which was created through the instrumentalization of nature for urban growth, is maintained today by a variety of contemporary institutions and systems that are neither just nor sustainable. Farming subsidies, international agreements, and transportation networks continue to ensure that food is grown in large monocultures and shipped all over the world. Water infrastructure continues to divert and channel Colorado River water to Los Angeles. Power lines and the organization of utilities continue to have centralized generation and long-distance transmission.

The political hegemony of the modern city becomes particularly significant in this ecological context. In order to reduce emissions, to decarbonize the grid, to electrify everything, particularly in the affluent cities of North America and Western Europe—those same cities created through the domestic and international imperialism of the previous centuries—decisions have to be made about the use of rural land, as well as wind, sun, water, and mineral resources. Decarbonizing the energy grid or conserving water requires coordinated changes in land use, infrastructure, and the built environment across country and city. Just like marshalling water for the growth of Los Angeles, decarbonizing energy systems for urban consumption involves decisions about rural land use and resources. And just as in the case of the construction of power lines in the American Southwest, decisions about urban futures are also decisions about rural ones. Thus, it is essential that these historic political and ecological relations are brought into focus.

Decarbonizing urban lives and livelihoods implies truly massive transformations of rural space in a new round of infrastructure development comparable in scale to that of the last century's, and comparable in its orientation toward urban needs to the previous rounds of rural extraction described above. To take the example of energy: most wind and solar that will power cities will be sited in deserts (Kruitwagen et al. 2021), just as the water that supplies Los Angeles comes from outside. And, owing to wind and solar's lower power density, renewables are a particularly land use-intensive energy regime, one that will require ten to a thousand times more land area than fossil fuels (Smil 2015). The coming geography of renewables will be far more horizontally extensive than fossil fuels, and most of those future land use needs and developments will be in rural areas (Huber and McCarthy 2017).

Globally, there is an understanding that rural communities are being asked to sacrifice again in the context of climate adaptation. Not only because these communities are disproportionally vulnerable to climate changes, such as heat and sea-level rise, but also because they are sites of the energy, food, and water systems that must be transformed, as well as where new rounds of extraction and development will take place. For these reasons there's been a querying of the politics of climate adaptation and an identification of ghosts of developmentalist discourses in contemporary discussions. Scholarship on "land grabbing" and "energy dispossession" in the context of climate change has flourished in the Global South (e.g., Baka 2017), where climate adaptation is producing a new round of financialization and profit-making based on urban needs and power. Geographer and development sociologist Kasia Paprocki (2020) has written of the "active erasure—social, epistemological, and material—of rural space and its alternative political imaginaries" in the context of urban climate planning, and its failure to imagine just rural climate futures (251).

Such analyses have been surprisingly absent in the United States, where a starkly and problematically bifurcated view of these transformations has dominated scholarship, politics, and media. Despite the physical proximity of rural energy, water, and food landscapes to urban areas, the "metronormativity" (Wang 2020; Buck 2021) of American visions of sustainability is striking. Rural issues of land use related to energy generation and siting are infrequently connected to sites of urban energy use and consumption in media, activism, or scholarship. Land use decisions remain nearly invisible in the cities where most media is produced and policy decisions are made, and they are poorly understood when they do appear. Policymakers are frustrated with opposition to renewable energy; urban climate movements are absent from discussions of rural land; scholars address urban energy consumption while eliding issues of rural generation. In short, there is no coherent public conversation around, or even awareness of, these interconnections at a moment when such significant decisions are being made.

This results in misunderstanding of rural protest and lost opportunities for coalition building. For instance, in the 2020s rural resistance to large-scale solar in the Great Basin and Mojave Desert has been understood as NIMBYism and climate denialism. Such forms of rural protest are most often discussed in mainstream media as regressive and anti-democratic, as examples of people putting private interests above public goods. But such characterizations miss the whole last century and a half of patterned decision-making, of the subservience of rural to urban needs. Such protest must be considered against the background of recurrent marshalling of resources for urban populations, of nature instrumentalized for urban growth, of what appear to be the rural costs of urban lifestyles. My own analysis has found protestors opposing a new wave of desert extraction rather than climate action per se, and articulating a progressive critique of capitalism and settler colonialism, not resisting adaptation (Angelo 2023).

In some ways, rural populations have a lot of political power in the United States, owing to their disproportionate electoral strength—and this has become increasingly challenging with the swell of right-populist discontent in rural America (I write this at the beginning of a second Trump term). But this belies an overall relative powerlessness of rural populations in decision-making about land and resource use. Geographically, American cities remain the literal seats of power. Federal policy, including over public lands based in Western states, is designed by institutions and policymakers based in Washington, DC, by people sitting in offices looking at maps. Nonprofits are for the most part based in large, coastal cities, and their perspectives on climate problems and solutions reflect those priorities and life experiences. Researchers and other knowledge producers, based at R1 institutions often in urban areas, provide the perspectives on rural land that have the biggest role in shaping decisions.

In short, cutting-edge visions of sustainable cities are premised on food, energy, building materials, and the disposal of waste in hinterlands, while the rural people who oppose such visions are depicted as backward, as climate deniers, as resistant to change, as unwilling to sacrifice, and/or as unable to see the reasonableness of these suggestions, given the necessity of these transformations. The urban imperialism of the nineteenth and twentieth centuries is being repeated in both the material relationships that are being proposed, in the infrastructure of climate adaptation, and the public discourse surrounding it.

CONCLUSION

Climate change is many things. Among them, it is an opportunity to change sedimented environmental and metabolic relationships that have proved themselves unsustainable as well as unjust.

Thus the questions arise: What would it take to dismantle these material and intellectual relations, in remaking infrastructure, built form, and ecological systems? To make decisions about resource use that respected both country and city, in the interests of justice and democracy as well as sustainability? Can we envision a pursuit of sustainability that extends beyond city boundaries, that has an eye focused both on urban consumption and rural needs (as well as nonhuman environmental needs, such as biodiversity loss and extinction crisis), and beyond urban imaginaries? And what kinds of scholarship might help us get there?

There's not just one answer to these questions. However, in trying to address them two things seem clear.

A first requirement is to have view of material interconnections and relations between city and hinterland at the forefront that acknowledges a basic historical pattern of rural extraction and urban accumulation in these dynamics. Important efforts to establish the "socioecological" essence of cities, in the context of the persistent amnesia of cities' reliance on their hinterlands, have sidelined or even effaced this older argument about cities' political and material imperialism: their ability to pull things into their orbits, and the unequal power relations between cities and their outsides. These political, spatial, and material dynamics remain relevant, and they need to be brought back to the fore in this moment of rethinking the systems on which (predominantly urban) life depends.

The second requirement is to understand that the metabolic and political pattern of urban imperialism that has dominated nineteenth- and twentieth-century life is deeply political, both in terms of who decides and what the trajectories of social change are. Thus, a historical view of who has paid the costs of consumption, whether costs have been distributed fairly, and if the same people must continue to pay—or if it is someone else's "turn"—seems essential as we attempt to understand current conflicts and discourses around adaptation and make decisions about the distribution of costs and benefits today. So we also need a view of the politics of urban-rural and urban-environmental relations that is sensitized to links between places and understandings of the land and its needs and uses from each location. Only then can we make decisions that are both democratic and sustainable, regarding, for instance, whether adaptation results in further impoverishment of the countryside or a more balanced set of costs and tradeoffs.

REFERENCES

Angelo, Hillary. 2017. "From the City Lens toward Urbanisation as a Way of Seeing: Country/City Binaries on an Urbanising Planet." *Urban Studies* 54 (1): 158–78.

Angelo, Hillary. 2021. *How Green Became Good: Urbanized Nature and the Making of Cities and Citizens*. University of Chicago Press.

Angelo, Hillary. 2023. "Boomtown: A Solar Land Rush in the West." *Harper's Magazine*, January.

Angelo, Hillary, and Miriam Greenberg. 2023. "Environmentalizing Urban Sociology." *City & Community* 22 (4): 257–65.

Angelo, Hillary, and David Wachsmuth. 2020. "Why Does Everyone Think Cities Can Save the Planet?" *Urban Studies* 57 (11): 2201–21.

Arboleda, Martín. 2020. *Planetary Mine: Territories of Extraction under Late Capitalism*. Verso.

Baka, Jennifer. 2017. "Making Space for Energy: Wasteland Development, Enclosures, and Energy Dispossessions." *Antipode* 49 (4): 977–96.

Brechin, Gray. 2000. "Author's Response to Critics." *Antipode* 32 (4): 513–17.

Brechin, Gray. 2006. *Imperial San Francisco: Urban Power, Earthly Ruin*. University of California Press.

Brenner, Neil, and Christian Schmid. 2015. "Towards a New Epistemology of the Urban?" *City* 19 (2–3): 151–82.

Buck, Holly Jean. 2021. *Ending Fossil Fuels: Why Net Zero Is Not Enough.* Verso.

Castree, Noel. 2000. "Between Mumford and Marx: Anatomising Imperial San Francisco." *Antipode* 32 (4): 494–99.

Cronon, William. 1991. *Nature's Metropolis: Chicago and the Great West.* W.W. Norton.

Finney, Carolyn. 2014. *Black Faces, White Spaces: Reimagining the Relationship of African Americans to the Great Outdoors.* UNC Press.

Federici, Sylvia. 2004. *Caliban and the Witch: Women, the Body, and Primitive Accumulation.* Autonomedia.

Foster, John Bellamy. 1999. "Marx's Theory of Metabolic Rift: Classical Foundations for Environmental Sociology." *American Journal of Sociology* 105 (2): 366–405.

Gandy, Matthew. 2003. *Concrete and Clay: Reworking Nature in New York City.* MIT Press.

Grove, Richard H. 1996. *Green Imperialism: Colonial Expansion, Tropical Island Edens and the Origins of Environmentalism, 1600–1860.* Cambridge University Press.

Harvey, David. 1996. *Justice, Nature and the Geography of Difference.* Blackwell.

Heynen, Nik, Maria Kaika, and Eric Swyngedouw, eds. 2006. *In the Nature of Cities: Urban Political Ecology and the Politics of Urban Metabolism.* Routledge.

Holleman, Hannah. 2018. *Dust Bowls of Empire.* Yale University Press.

Huber, Matthew T., and James McCarthy. 2017. "Beyond the Subterranean Energy Regime? Fuel, Land Use and the Production of Space." *Transactions of the Institute of British Geographers* 42 (4): 655–68.

Kahrl, William L. 1982. *Water and Power: The Conflict over Los Angeles Water Supply in the Owens Valley.* University of California Press.

Kaika, Maria. 2004. *City of Flows: Modernity, Nature, and the City.* Routledge.

Kruitwagen, L., Story, K.T., Friedrich, J., Byers, L., Skillman, S., & Hepburn, C. 2021. "A Global Inventory of Photovoltaic Solar Energy Generating Units." *Nature* 598 (7882): 604–10.

Logan, John R., and Harvey Molotch. 2007. *Urban Fortunes: The Political Economy of Place.* University of California Press.

Moore, Jason W. 2015. *Capitalism in the Web of Life: Ecology and the Accumulation of Capital.* Verso.

Needham, Andrew. 2014. *Power Lines: Phoenix and the Making of the Modern Southwest.* Princeton University Press.

Paprocki, Kasia. 2020. "The Climate Change of Your Desires: Climate Migration and Imaginaries of Urban and Rural Climate Futures." *Environment and Planning D: Society and Space* 38 (2): 248–66.

Polanyi, Karl. 2001. *The Great Transformation: The Political and Economic Origins of our Time.* Beacon Press.

Reisner, Marc. 1993. *Cadillac Desert: The American West and Its Disappearing Water*. Penguin.

Robinson, Jennifer. 2002. "Global and World Cities: A View from off the Map." *International Journal of Urban and Regional Research* 26 (3): 531–54.

Smil, Vaclav. 2015. *Power Density: A Key to Understanding Energy Sources and Uses*. MIT Press.

Taylor, Dorceta E. 2016. *The Rise of the American Conservation Movement: Power, Privilege, and Environmental Protection*. Duke University Press.

Thompson, E.P. 1963. *The Making of the English Working Class*. Vintage.

Wang, Xiaowei. 2020. *Blockchain Chicken Farm: And Other Stories of Tech in China's Countryside*. Farrar, Straus and Giroux.

Water Education Foundation. 2023. https://www.watereducation.org/.

Wittfogel, Karl. 1957. *Oriental Despotism: A Comparative Study of Total Power*. Yale University Press.

Worster, Donald. 1982. "Hydraulic Society in California: An Ecological Interpretation." *Agricultural History* 56 (3): 503–15.

Worster, Donald. 1992. *Rivers of Empire: Water, Aridity, and the Growth of the American West*. Oxford University Press.

14. From Crisis to Countermovements

Diana Stuart

Scientists continue to stress that we face a serious environmental crisis. This crisis includes multiple related crises: climate change, biodiversity loss, and the crossing of an increasing number of critical biophysical thresholds (e.g., see Steffen et al. 2018; Ripple et al. 2020). The projected impacts could threaten all of humanity and many scientists claim that the extent of these impacts is likely underestimated.

Here, I focus on the climate crisis, acknowledging that it remains part of the larger environmental crisis driven by humans. Average global temperatures are increasing each year, which shifts climate dynamics and causes unprecedented heat, drought, fires, floods, hurricanes, sea level rise, and biodiversity loss, among other harmful and hazardous impacts. Because relatively little has been done to mitigate global warming, we are on a trajectory to easily surpass the Paris Agreement target of staying within 1.5°C of warming. According to Climate Action Tracker (2023), current policies will result in *at least* 2.6°C of warming by 2100 with an increase of 1.8°C being the most optimistic outcome if there is a rapid surge in radical action. We face a planetary emergency that global leaders should have started addressing decades ago. This late in the game, what can be done? Is it too late to act?

It is not uncommon to encounter individuals who believe "we are doomed" and there is no chance of escaping climate-driven apocalypse. This trend of more and more people feeling doomed is widely referred to as **doomerism**. Climate change deniers have now been largely replaced by climate "doomers" who believe it is no longer worth trying to mitigate climate change (Osaka 2023). Are they right? While there are biophysical thresholds that can result in irreversible and catastrophic changes, scientists cannot know for certain when these thresholds might be reached. What is certain is that a 2°C warmer future will be more livable than a 3°C or 4°C warmer future. As David Wallace-Wells (2019) explains in *The Uninhabitable Earth*, "global warming is not binary. It is not a matter of 'yes' or 'no,' not a question of 'fucked' or 'not' . . . it is a problem that gets worse over time." The science confirms this. For example, in terms of climate-related mortality, Daniel Bressler (2021) conservatively estimates that below 2°C the estimated annual excess deaths related to extreme temperatures (heat or cold alone) are around 100,000, but above 2°C the estimated associated deaths increase, reaching more than four million at 4°C. Stated in different terms, every 4,434 metric tons of carbon added beyond the 2020 rate of emissions kill one person and, at our current trajectory, we would see eighty-three million excess deaths related to climate change by 2100 (Bressler 2021). Every fraction of additional warming will have more severe impacts, causing more suffering, loss, and death.

What this means is that every degree matters. Believing that once we pass 1.5°C we are doomed and should therefore give up illustrates a misguided understanding of the climate crisis. As reporter Shannon Osaka (2023) explains, much of this "doomer" thinking is based on media reports of the 2018 IPCC special report on 1.5°C, which focused on a twelve-year target to cut emissions. This was widely interpreted as "if we don't stop climate change in 12 years, something catastrophic will happen." Demoralized by a lack of any meaningful responses to these warnings, many people now believe we are doomed. Yet, prematurely accepting doom undermines action and the potential for a more livable future. Giving up now will only result in more warming, suffering, and loss. Doomerism may very well have emerged organically owing to increasingly dire scientific reports and a complete lack of adequate responses; however, it leads to inaction and maintaining the status quo and therefore is very likely supported or even propagated by those invested in fossil fuels. If we stop trying to avoid catastrophic warming, things are made much easier for them. Understanding and countering doomism has, therefore, become a critical step in galvanizing the social responses necessary to avoid the worst of climate change.

Doomerism relates to defeatism and fatalism. Defeatism is exemplified by the novelist Jonathan Franzen (2019), who in a *New Yorker* article titled "What if We Stopped Pretending?" claims that an "all-out war on climate change made sense only as long as it was winnable." Defeatists like Franzen believe that we have already lost or that the battle simply cannot be won. There is no point in acting, the logic goes, because one cannot succeed. Fatalism involves believing that the worst climate outcomes are inevitable: it is simply our fate. Therefore, there is no point in acting because the future is already determined.

With both defeatism and fatalism, all agency to change the situation is lost. "There is no belief in any action making a difference" (Lueck 2007, 251). As Franzen (2019) argues, we should just accept our defeat/fate and focus on what we can still enjoy: family, friends, and the things in life that give us pleasure and provide happiness. However, this is a very privileged position. While wealthy people may have the means to attempt to avoid the worst impacts of climate change, the global majority will not be able to enjoy their lives when faced with life-threatening crises like famine, fires, and floods.

Given the warming projections and the continued absence of meaningful responses, pessimism is warranted. However, a pessimistic view need not be defeatist or fatalistic. We are already experiencing global warming and a non-warming future is no longer possible. Currently, there is also little evidence of a rapid "World War II-level mobilization" to prevent as much future warming as possible. Yet this does not have to result in despair in terms of inaction. Despair and inaction on climate change are simply "a self-fulfilling prophecy" (Lueck 2007, 251). It is critical to debunk doomerism and to counter defeatism and fatalism. There is still much that can be done to reduce the extent of warming. There are also alternatives that most people have failed to consider. As Ryan Gunderson (2020) puts it, even when the situation is far from encouraging, "a pessimistic perspective ought to avoid the dusk of fatalism through a persistent search for alternatives" (610). It is this search for alternatives that can result in a path forward. We cannot stop global warming, but we can still reorganize and reconstruct our social system to significantly reduce the extent of warming and the associated suffering and loss. In other words, there are still possibilities to create a less warm and a more livable future.

Environmental sociologists and others have taken key steps toward understanding what system change to justly minimize warming might entail and how it might come to fruition. While there are many obstacles and challenges preventing system change, it remains a worthy endeavor to examine, discuss, and push forward to develop possibilities for creating a more livable future. In this chapter, I examine what could be done to minimize global warming. Calls for system change have increased, but what does this mean? I discuss what system change would need to entail to fairly minimize climate impacts. Next, I discuss how system change could occur and what social transformation might look like. I then briefly examine the climate movement—what it has done and what it might need to do to push forward a transformative climate agenda. Lastly, drawing from moral philosophy, I provide reasons to act and to keep acting to minimize warming, no matter what. Even though global warming is happening and will continue to happen, there is still much that can be done to lessen the extent of its impacts.

WHAT KIND OF SYSTEM CHANGE?

Following the Intergovernmental Panel on Climate Change's (IPCC) (2018) *Special Report on 1.5 Degrees*, which stated that "rapid and far-reaching" changes are necessary in all aspects of society, calls for "system change" to address global warming became more and more common. In 2019 and 2020, calls for system change, especially "System Change, Not Climate Change," were commonly written on protest signs. In 2021, the "code red for humanity" IPCC report further amplified calls for system change. Yet, what does system change mean? What would system change that is aimed at justly minimizing warming entail?

Despite the propaganda, system change entails much more than individual, consumer-based change. The estimates regarding the potential contributions of individual behavior changes vary, but all confirm that the majority of emissions come from companies and states, not individuals (Heede 2014). A few examples illustrate this clearly. In the United States, the commercial and industrial sectors use the majority of energy, with residential use only representing 34 percent of total energy use (EIA 2018). In the European Union, a *complete* shift to green consumerism is estimated to reduce emissions by only 25 percent (Moran et al. 2018). Another study estimates that the widespread adoption of thirty different behavioral changes could mitigate 19 to 36 percent of global emissions (Williamson et al. 2018). While the initial reduction in carbon emissions due to the COVID-19 pandemic was around 17 percent, one scientist explained that "at the same time, 83% of global emissions are left, which shows how difficult it is to reduce emissions with changes in behaviour . . . Just behavioural change is not enough" (Harvey 2020).

Despite all this, there are still very good reasons to personally reduce our carbon footprint. These include: (1) a personal refusal to be complicit in additional warming and related harm, and (2) actively demonstrating that low-carbon living is possible and could become the new norm. Yet we cannot stop with individual change, as it is not even close to enough to avoid climate catastrophe. The majority of emissions remain out of our reach and therefore require much larger systemic changes.

System change must also go beyond simply adding more renewable energy to our current system. The idea that renewable energy development alone will reduce emissions relies on the assumption that renewable energy production will displace fossil fuels. Yet empirical analyses suggest a true energy "transition" is not occurring in the way that many people assume it is. This is because expanding renewable energy does not necessarily displace fossil fuels so much as it is adding to increases in total energy production and consumption (Zehner 2012; York 2016; York and Bell 2019). Ryan Thombs (2017) has coined the term "renewable energy paradox" to describe the counterintuitive outcome that renewable energy has little influence on carbon emissions in developed countries. In a system with continued economic growth, we see consistent increases in production, consumption, and energy. This means that as total energy use continues to grow, renewables are largely being added to the energy system rather than replacing fossil fuels. Renewable energy development in current conditions, according to Thombs, York, and others, merely creates additional capacity for production. In a system where energy use per capita keeps going up, renewable energy remains limited to marginal emissions reductions (Adua et al. 2021).

Why are these popular "solutions" not enough? While fossil fuels are often identified as the driver of climate change, there lies a deeper and more systemic root driver. For example, there are other sources of emissions (adding up to 30 percent), which include deforestation, animal agriculture, and synthetic fertilizer use. In a system that is no longer powered by fossil fuels, continued expansion in these other production systems would continue to contribute to global warming (see Hickel 2016). Increasing evidence suggests that the root driver of this crisis is the growing levels of global production and consumption, which are increasing per capita each year (see Kallis 2018; Hickel and Kallis 2019; Schor and Jorgenson 2019; Hickel 2020). While it is certain that some people still need more resources for survival and well-being, more and more people (especially in wealthy countries) continue to unnecessarily produce and consume simply for the sake of growth. This is not so much a problem of individual consumption choices, but of a production system that uses advertising and other means to create the consumption necessary to sustain profitable levels of ever-increasing production (Wiedmann et al. 2020).

Indeed, there is a clear positive association between carbon emissions and economic growth. Growth is usually measured by Gross Domestic Product (GDP), or the total market value of all goods and services produced. GDP in the United States has increased on average 3 percent each year since World War II. In his famous 2006 review, Nicholas Stern recognized an important trend: reduced economic growth associated with recessions resulted in a total decrease in carbon emissions. This trend has continued. Data increasingly shows a positive relationship between GDP growth and emissions (Stern

2006; Jorgenson and Clark 2012; Burke et al. 2015; Hickel and Kallis 2019). As a result, more environmental sociologists and other scholars and activists are stating that there is a clear incompatibility between continued economic growth and effectively addressing climate change (e.g., Hickel and Kallis 2019; Schor and Jorgenson 2019). As stated by William Ripple and colleagues (2019, 4), "our goals need to shift from GDP growth and the pursuit of affluence toward sustaining ecosystems and improving human well-being."

What about the idea of **green growth**, the "win-win" strategy where technology can be used to decouple economic growth from environmental impacts? The viability of the case for green growth requires absolute reductions in emissions despite economic growth (absolute decoupling). While evidence of decoupling depends on what is measured and over what time, a recent review found that "even countries that have achieved absolute decoupling are still adding emissions to the atmosphere thus showing the limits of 'green growth' and the growth paradigm" (Hubacek et al. 2021). In another study, Helmut Haberl and colleagues (2020) state that, while absolute decoupling of both production and consumption-based emissions can be found in some cases, it is not occurring globally or at the rates necessary; it therefore "needs to be complemented by sufficiency-oriented strategies and strict enforcement of absolute reduction targets." It is not that transformations in energy and technology are misguided, but that technological changes are insufficient to meet climate targets (Schor and Jorgenson 2019). Growth continues to undermine technological mitigation tools and the prospects for a livable future (Stuart et al. 2020).

Therefore, system change—to justly minimize global warming or even avoid the worst impacts of warming—must curtail economic growth. Rather than simply changing consumption patterns or moving toward a fossil fuel-free system, we also need to reduce economic growth in wealthy countries (called degrowth—see Kallis 2018; Hickel 2020). In other words, we must change the entire growth-focused system. This requires transitioning to a material system of **sufficiency**: having enough, rather than always producing and consuming more and more. This level of system change is possible but it requires an understanding of how system change occurs and what specifically needs to be changed to transition to a society that can minimize warming.

HOW COULD SYSTEM CHANGE OCCUR?

As stated in the previous chapter, climate change is many things; it includes an opportunity to remake our current system in ways that result in more sustainable and more just outcomes. Many scholars in recent years have looked to the work of Erik Olin Wright to understand current possibilities for system change. In *Envisioning Real Utopias,* Wright (2010) uses historical patterns to articulate a theory of social transformation that includes: (1) identifying and challenging forces of social reproduction that maintain the status quo; (2) creating and politicizing problems with the current social order (eroding its legitimacy); (3) simultaneously adopting multiple transformational strategies that involve creating alternatives/"building the new," as well as eroding the old through a series of reforms; (4) pressuring leaders to direct change through social disruption or the "logic of rupture" (see also Wright 2019). Here, I focus on how these strategies are being used, or could be used, to catalyze system change.

As Wright (2010) explains, "social structures and institutions that systematically impose harms on people require vigorous mechanisms of active social reproduction in order to be sustained over time" (276). Social change requires identifying these forces that maintain the status quo (often called forces of social reproduction). It also requires understanding their strategies and then exposing, challenging, and delegitimizing them. Environmental sociologists have made significant contributions to identifying and publicly exposing actors promoting climate change denial and climate change obstruction. For example, Aaron McCright and Riley Dunlap's (2000, 2003, 2010, 2011) work on climate denial exposed the perpetrators of denial and other tactics to undermine climate policy at the international and national levels. This work was highlighted in many major news outlets. Another example is Robert Brulle's work (2014, 2021, 2023) on climate obstruction and the **climate change countermovement**, which highlights how vested interests use a range of tactics to purposefully delay and undermine climate policy. This work was also featured in major news outlets and continues to be used to expose those obstructing climate action. Exposing forces of social reproduction and their strategies will continue to illustrate how these actors are undermining what is best for humanity for their own personal gain.

Work is also already being done to erode the legitimacy of the actors and institutions that actively maintain the status quo. Sugandha Srivatav and Ryan Rafaty (2021) highlight the use of delegitimating strategies and "antagonistic" actions. These can include directed protests, sit-ins, naming and shaming tactics, consumer boycotts, or divestment campaigns. This can influence consumption and voting behavior; more importantly, it can be used as part of a narrative, along with the politicization of specific events, to cast doubt on leaders failing to act on climate change. Lawsuits are also a mechanism to publicly blame and shame companies and force government action. As Srivtav and Rafaty (2021) explain, "the antagonist mantra can be summarized by: name, shame, boycott, and sue"(11). In addition, framing climate change as a moral issue centered on intergenerational injustice may be a key leverage point to delegitimize those who maintain the status quo. Wright (2019) argues that most people are motivated by moral concerns rather than class or economic concerns. As youth activists increasingly demand that world leaders protect their future, they draw attention to the immorality of continuing with business as usual. All this helps to undermine faith in the current system and casts doubt on the actors and institutions who have been upholding the system for their own benefit.

In addition to weakening the position of those in power, new policies, programs, and institution need to be created to build a new system as the previous one erodes. As Wright (2010) shows, this process typically happens slowly through "synergistic" strategies, which are based on class compromise and incremental policy changes. However, the character of climate change leaves little time for compromise and incremental change. There is a growing focus on the potential for **non-reformist reforms** or policies that change key systemic levers and open the door for further social transformation (Gorz 1967; Kallis 2018; Stuart et al. 2020). Proposals for a "Green New Deal" could be transformative, yet most fail to include system-changing strategies that would represent non-reformist reforms. Faced with compromise ahead, those proposing climate policies and programs will need to aim high and demand what may seem radical in order to end up with meaningful mitigation policies.

Pushing forward just and effective climate action in the current context would require immense public pressure and the ability to challenge power. Giorgos Kallis (2018) explains that "expecting that the dominant classes will somehow release their power and forgo their immediate interests in the name of a broader common good is unrealistic" (141). Wright's (2010) ideas on the "logic of rupture" would support using strategic disruption through non-violent civil disobedience to pressure leaders to take bold action. Social movements do have the potential to pressure governments, elites, and international governing bodies to adopt new policy platforms and goals (Almeida and Chase-Dunn 2018). However, Frances Fox Piven (2008) argues that effectively challenging power requires developing networks of solidarity that make collective action a meaningful leverage point. To challenge power, a social movement must be large and united. Piven describes how effectively challenging power requires masses of people ceasing to cooperate and withdrawing participation in the system to the extent that the rest of the system cannot function. This level of unity is difficult to achieve, but it is highly effective.

A CHALLENGE FOR THE CLIMATE MOVEMENT

All this is highly important to consider as the climate movement continues to grow, explores new strategies, and begins to more clearly articulate specific demands and policy platforms. While an ephemeral movement existed previously and was seen in events such as the 2014 People's Climate March in New York City, a larger and sustained climate movement has emerged more recently. Between 2018 and 2020, the climate movement grew to an unprecedented size with large-scale global protests demanding government action. Multiple climate movement organizations have taken the spotlight using a variety of tactics and strategies.

Fridays for Future (FFF) was inspired by Greta Thunberg, who at the age of fifteen, was protesting alone outside the Swedish parliament every Friday, demanding action to address the climate and ecological crisis. This ultimately resulted in the youth-led group Fridays for Future carrying out "school strikes for climate" on Fridays. Fridays for Future gained increased attention in 2019: approximately 1.5 million students participated in a global strike in March 2019 and around six million people participated in two consecutive general strikes in September 2019 (Taylor et al. 2019). FFF represents a powerful force to frame climate change in moral terms, bringing attention to the many children whose futures will be negatively impacted by climate change.

Extinction Rebellion (XR) initiated a wave of protests and acts of civil disobedience in the United Kingdom that spread internationally. Their strategy was to repeatedly shut down the center of government and commerce (London, e.g. until the government meets their three demands. These include the following: (1) telling the truth about the climate and ecological crisis; (2) reducing carbon emissions to net zero by 2025; and (3) creating a democratic citizens assembly to guide the transition. In November 2018, over six thousand activists shut down five major bridges in London (Taylor and Gale 2018). Before changing tactics, XR continued to shut down central areas of London and other cities in the United Kingdom during annual fall and spring "rebellions."

The Sunrise Movement works through the channels of US electoral politics to strategically get people elected who will support climate action and to advocate for climate policy, even if it is reformist rather than revolutionary. They have, by most accounts, changed the climate discussion in the

United States and helped to push forward the first piece of national climate legislation in the nation's history. While hailed by some as a victory, as it provides $369 billion for energy and climate change related projects, the 2022 Inflation Reduction Act was a clear compromise that fails to even end fossil fuel subsidies. However, many claim that it represents an important first step and the Sunrise Movement continues to fight for stronger climate mitigation policies.

Despite a temporary stagnation in climate activism caused by the global COVID-19 pandemic, climate activist groups continue to organize direct actions, demanding bold climate responses from world leaders. XR and FFF continue to hold seasonal rebellions and school strikes on Fridays, respectively. Other groups have increased direct actions, such as Ende Gelände blocking coal mines in Germany and Just Stop Oil delaying the transport of fossil fuels in the UK. There is no doubt that a climate movement has emerged that is larger, more diverse, and growing more rapidly than ever. Yet we do not see a bold and rapid response. Many scholars and activists cite the work of Stephan and Chenoweth (2011), who claim that when a critical mass of 3.5% of the population participates in sustained non-violent disruption it can trigger large-scale social transformation. However, at this level of engagement, there are few examples of sustained activism beyond movements in response to repressive and autocratic leadership (Fisher 2022). New strategies are necessary to increase the size and strength of the climate movement. It is also possible that unexpected events may trigger a surge in climate activism and catalyze the levels of participation necessary to challenge power.

As André Gorz (1967) pointed out, it is one thing to want to change the system, yet it is another thing to know what specifically to do once you have the power to do so. In other words, we need to know what specifically is needed to create a new system or in this case as system that can justly minimize warming. Within the groups calling for system change, a large gap remains between the complaints and discontent with the current system and any agenda that would result in meaningful change (Kenis and Mathijs 2014; Stoner and Melathopoulos 2016). As Clive Spash (2020) explains, these "generalized complaints" about the failures of the current system remain unspecific and therefore the agenda of climate activists remains "disconnected and incomplete." Alternatives to the dominant paradigm of "progress" will remain marginalized and ignored if these alternatives continue to be generic calls for "system change" without any articulation of what this must involve. As Spash (2020) points out, these generic calls for change and action fail to connect with any agenda or program to facilitate change and also critically fail to confront the very real political powers that continue to protect the status quo.

These challenges stem in part from a societal inability to understand that different socio-economic systems can work and are possible to create through specific strategies. Trapped in **capitalist realism** (Fisher 2009), the majority of people remain blinded to the reality that there are alternatives to capitalism. However, how can society break free of capitalist realism if alternatives are not fully articulated or understood in concrete and tangible ways? Calls for system change must get louder, but they must also be much more specific. An understanding of specific structural policies to create a new system remains a crucial, yet missing, element in the climate movement.

Luckily, much research has already been done using empirical data to identify the most promising strategies to justly minimize warming. Drawing from the work of degrowth and ecosocialist scholars, my coauthors and I (Stuart et al. 2023) outline one possible agenda to guide system change. This agenda includes widespread economic democracy to open up the possibilities for firms to have priorities besides growth; work-time reduction to curtail overproduction- and production-related energy use and carbon emissions; energy democracy and cooperatives to simultaneously meet emissions reduction targets and social needs; advertising restrictions to curtail harmful consumption; reallocation/redistribution of excess wealth (which is linked to extreme carbon emissions); and nationalizing and phasing out fossil fuel companies. These specific non-reformist reforms could be implemented independently but would have the greatest impact if implemented in concert. In addition, they could be promoted without using potentially misunderstood or divisive terms like ecosocialism and degrowth. Having a specific agenda with concrete policies is key for the climate movement to effectively catalyze system change. The challenge remains connecting these types of ideas to the climate movement in a way that can galvanize support and make the movement more powerful.

Currently, it is highly unlikely that global leaders will adopt the above-described policies and programs. However, given the potential of unexpected events to open up some of the routes for change described above, it is not impossible. It is clearly worth imagining, articulating, sharing, and proposing such ideas. In addition, asking and demanding the most "radical" actions to minimize warming is more likely to result in the greatest level of mitigation to emerge after negotiation and compromise, the usual path for synergistic change (Wright 2010). Non-reformist reforms will need to go through the political channels and to be as effective as possible proponents will need to aim high and resist compromise. While the odds of a radical agenda being implemented is currently low, it is worth pursuing for multiple reasons: (1) aiming high is a smart political strategy and (2) we have a moral obligation to try to minimize as much climate-related harm as possible. As climate impacts translate into suffering, loss, and death, who can say that it is acceptable to allow any preventable warming to occur?

THE OBLIGATION AND WILL TO ACT

As Erich Fromm (1976) explained years ago, the environmental crisis is a life-or-death situation. Fromm compares the crisis to a hospital patient who will die unless they receive a specific procedure from the doctors on duty. Even if the chance of the procedure succeeding is extremely low, Fromm argues that the doctors have a moral obligation to perform the procedure. It is their duty to act to preserve life, even if there is only a small chance that their actions will be successful. In the context of the environmental crisis, he argues, we are all obligated to act to preserve life despite the immense challenges and low odds of success. Taking this analogy further, we can imagine that the patient has no chance of a full recovery that results in resuming their life as exactly before. Changes have occurred that are irreversible and there are likely more changes that will be revealed if the patient lives. Yet, despite what may already be lost and the low odds of a successful operation, it is still the duty of the doctors to act to preserve life. They must act despite what is already lost and despite the slim chances of success.

Global warming is already occurring and will continue to occur with known and unknown impacts. Yet action can still be taken to preserve life. A full return to what was before is not possible, but we can work toward what is still possible: a less warm and more livable future than our current trajectory. There is still time and ample ability to act to reduce the extent of impacts and save many lives. We may grieve what has already been lost and what will be lost, but with so much left to save, it is morally obligatory to act. While moral philosophers may argue about many things, even when drawing from differing ethical schools of thought, the right thing to do is to act to minimize climate-related loss and suffering (Garvey 2008). In addition, based on any understanding of fairness, equity, or justice, those who have contributed the most to carbon emissions and benefited the most from past emissions should be shouldering the greatest burden in mitigating the climate crisis and aiding in adaptation efforts. That means people in wealthy countries are obligated to aggressively cut emissions and provide assistance to those who lack the resources to adapt (Cripps 2022). In short, there is a moral obligation to act, especially for those of us living in countries most responsible for this crisis.

Harnessing the will to act in these discouraging circumstances relates to the notion of radical hope. Radical hope arises when the best-case scenario is no longer possible. In this case, hoping for a non-warming and rosy future is pure delusion. In addition, much will be lost. But, as Thompson (2010) explains, radical hope is "the hope for revival: for coming back to life in a form that is not yet intelligible" (48). With so many people succumbing to defeatism or fatalism when faced with the realities of the climate crisis, cultivating radical hope remains a key task moving forward. Radical hope is authentic in that it acknowledges the truth about loss and the low odds of success but demands action because there is something left to be saved or at least the possibility that something might emerge in place of what was (Lear 2006; Thompson 2010; Williston 2015). Radical hope is continued action with the belief that something good will be left or that something good will materialize even when all else is lost. This belief in the possibility of a future good can sustain continued action despite the extent of loss.

Virtue ethics, in Aristotelian terms, can also serve to fuel sustained actions to demand bold climate action and in many ways is highly suited to the challenges of the climate crisis (Morrell and Dalmann 2022). This involves a focus on the agents, their character and their actions, despite the outcomes. Byron Williston (2015) discusses the importance of three specific virtues for climate activists: justice, truth, and hope. These involve justice for future generations and species, telling the truth about the reality of the situation and rejecting false optimism and technological fantasies, and an active hope for what is still possible. As summarized by Benjamin Lowe (2019), Williston argues that an advantage of a virtue ethics approach for climate activism is that it enables "moral agents to do what is right even when the results are not likely to be particularly effective or rewarding. Thus, they are more dependent on our character than on our perceived prospects for success" (481–82). This approach is focused on a commitment to action and doing the right thing despite the odds of achieving the desired outcome.

Action can take on a variety of forms. Although it is not a standalone solution, this can include lifestyle changes that reduce personal emissions and illustrate a refusal to be complicit in the harmful system. However, changing our individual lifestyle choices should not mean that we overlook the absolute necessity of demanding larger system change. There are many ways to work toward system change, including voting, donating, volunteering, running for office, communicating alternatives, encouraging action, organizing action, protesting, and participating in civil disobedience. Environmental sociologists also continue to play a critical role in working towards system change to justly address climate change. This includes exposing those undermining climate policy and their tactics, studying the climate movement and what strategies might be most effective, studying system alternatives and specific policies that might most justly minimize warming, identifying pathways for change, participating in policy-formation and governmental science panels, and communicating their work through teaching, writing, news interviews, public talks, and opinion or editorial pieces. There is much to be done and many ways that we can all actively keep working towards change.

While many people feel lost and helpless due to the failure of global leaders to address the climate crisis, radical hope and agent-focused virtue ethics can be used to avoid despair and sustain action. Moving forward, it is critical we avoid despair. As Treanor (2010) explains:

> Despair is fatal to both environmental progress and individual flourishing . . . It is fatal to environmental progress because while it is true that we may not be able to adequately respond to certain crises in time to avoid their negative effects, failing to try ensures failure and often exacerbates the situation. Despair is fatal to flourishing because it undermines our belief in the significance of our actions and our lives. (26)

In other words, rejecting despair increases the possibilities for positive social-ecological outcomes and it can also support personal wellbeing. Our environmental crisis is indeed a life-or-death situation. Despite the immense challenges and now inevitable losses, the moral response is still to act and to continue to act, especially when there is still so much left to save.

REFERENCES

Adua, Lazarus, Karen Xuan Zhang, and Brett Clark. 2021. "Seeking a Handle on Climate Change: Examining the Comparative Effectiveness of Energy Efficiency Improvement and Renewable Energy Production in the United States." *Global Environmental Change* 70:102351.

Almeida, Paul, and Chris Chase-Dunn. 2018. "Globalization and Social Movements." *Annual Review of Sociology* 44:189–211.

Bressler, R. Daniel. 2021. "The Mortality Cost of Carbon." *Nature Communications* 12 (1): 4467.

Brulle, Robert J. 2014. "Institutionalizing Delay: Foundation Funding and the Creation of US Climate Change Counter-Movement Organizations." *Climatic Change* 122:681–94.

Brulle, Robert J. 2021. "Networks of Opposition: A Structural Analysis of US Climate Change Countermovement Coalitions 1989–2015." *Sociological Inquiry* 91 (3): 603–24.

Brulle, Robert J. 2023. "Advocating Inaction: A Historical Analysis of the Global Climate Coalition." *Environmental Politics* 32 (2): 185–206.

Burke, Paul J., Md Shahiduzzaman, and David I. Stern. 2015. "Carbon Dioxide Emissions in the Short Run: The Rate and Sources of Economic Growth Matter." *Global Environmental Change* 33:109–21.

Climate Action Tracker. 2023. https://climateactiontracker.org/global/temperatures/.

Cripps, Elizabeth. 2022. *What Climate Justice Means and Why We Should Care*. Bloomsbury.

EIA (U.S. Energy Information Administration). 2018. Electricity. eia.gov/electricity/annual/.

Franzen, Jonathan. 2019. "What if We Stopped Pretending?" *New Yorker*, September 8. https://www.newyorker.com/culture/cultural-comment/what-if-we-stopped-pretending.

Fromm, Erich. 1976. *To Have or to Be?* Harper & Row.

Garvey, James. 2008. *The Ethics of Climate Change: Right and Wrong in a Warming World*. Bloomsbury.

Gorz, André. 1967. *Strategy for Labor*. Beacon Press.

Fisher, Dana R. 2022. "AnthroShift in a Warming World." *Climate Action* 1 (1): 1–6.

Haberl, Helmut, Dominik Wiedenhofer, Doris Virág, Gerald Kalt, Barbara Plank, Paul Brockway, Tomer Fishman, et al. 2020. "A Systematic Review of the Evidence on Decoupling of GDP, Resource Use and GHG Emissions, Part II: Synthesizing the Insights." *Environmental Research Letters* 15 (6): 065003.

Harvey, Fiona. 2020. "Lockdowns Trigger Dramatic Fall in Global Carbon Emissions." *Guardian*, May 19. https://www.theguardian.com/environment/2020/may/19/lockdowns-trigger-dramatic-fall-global-carbon-emissions.

Heede, Richard. 2014. "Tracing Anthropogenic Carbon Dioxide and Methane Emissions to Fossil Fuel and Cement Producers, 1854–2010." *Climatic Change* 122 (1): 229–41.

Hickel, Jason. 2016. "Clean Energy Won't Save Us—Only a New Economic System Can." *Guardian*, July 15. https://www.theguardian.com/global-development-professionals-network/2016/jul/15/clean-energy-wont-save-us-economic-system-can.

Hickel, Jason. 2020. *Less Is More: How Degrowth Will Save the World*. Random House.

Hickel, Jason, and Giorgos Kallis. 2019. "Is Green Growth Possible?" *New Political Economy* 25 (4): 469–86.

Hubacek, Klaus, Xiangjie Chen, Kuishuang Feng, Thomas Wiedmann, and Yuli Shan. 2021. "Evidence of Decoupling Consumption-Based CO2 Emissions from Economic Growth." *Advances in Applied Energy* 4 (19): 100074.

IPCC. 2018. *Global Warming of 1.5°C. An IPCC Special Report on the Impacts of Global Warming of 1.5°C above Pre-Industrial Levels and Related Global Greenhouse Gas Emission Pathways, in the Context of Strengthening the Global Response to the Threat of Climate Change, Sustainable Development, and Efforts to Eradicate Poverty.* Edited by V. Masson-Delmotte, P. Zhai, H.-O. Pörtner, D. Roberts, J. Skea, P.R. Shukla, A. Pirani, et al. IPCC.

Jorgenson, Andrew K., and Brett Clark. 2012. "Are the Economy and the Environment Decoupling? A Comparative International Study, 1960–2005." *American Journal of Sociology* 118 (1): 1–44.

Kallis, Giogos. 2018. *Degrowth.* Agenda.

Kenis, Anneleen, and Erik Mathijs. 2014. "Climate Change and Post-Politics: Repoliticizing the Present by Imagining the Future?" *Geoforum* 52:148–56.

Lear, Jonathan. 2006. *Radical hope: Ethics in the Face of Cultural Devastation.* Harvard University Press.

Lowe, Benjamin S. 2019. "Ethics in the Anthropocene: Moral Responses to the Climate Crisis." *Journal of Agricultural and Environmental Ethics* 32: 479–85.

Lueck, Michelle A.M. 2007. "Hope for a Cause as Cause for Hope: The Need for Hope in Environmental Sociology." *American Sociologist* 38:250–261.

McCright, Aaron M., and Riley E. Dunlap. 2000. "Challenging Global Warming as a Social Problem: An Analysis of the Conservative Movement's Counter-Claims." *Social Problems* 47 (4): 499–522.

McCright, Aaron M., and Riley E. Dunlap. 2003. "Defeating Kyoto: The Conservative Movement's Impact on US Climate Change Policy." *Social Problems* 50 (3): 348–73.

McCright, Aaron M., and Riley E. Dunlap. 2010. "Anti-Reflexivity." *Theory, Culture & Society* 27:100–133.

McCright, Aaron M., and Riley E. Dunlap. 2011. "Cool Dudes: The Denial of Climate Change among Conservative White Males in the United States." *Global Environmental Change* 21 (4): 1163–72.

Moran, Daniel, Richard Wood, Edgar Hertwich, Kim Mattson, Joao F.D. Rodriguez, Karin Schanes, and John Barrett. 2020. "Quantifying the Potential for Consumer-Oriented Policy to Reduce European and Foreign Carbon Emissions." *Climate Policy* 20 (1): S28–S38.

Morrell, Kevin, and Frederik Dahlmann. 2022. "Aristotle in the Anthropocene: The Comparative Benefits of Aristotelian Virtue Ethics over Utilitarianism and Deontology." *Anthropocene Review* 10 (3): 615–35.

Osaka, Shannon. 2023. "Why Climate Doomers Are Replacing Climate Deniers." *Washington Post,* March 24. https://www.washingtonpost.com/climate-environment/2023/03/24/climate-doomers-ipcc-un-report/.

Piven, Frances Fox. 2008. "Can Power from Below Change the World?" *American Sociological Review* 73 (1): 1–14.

Ripple, William J., Christopher Wolf, Thomas M. Newsome, Phoebe Barnard, William R. Moomaw, and Philippe Grandcolas. 2020. "World Scientists' Warning of a Climate Emergency." *BioScience* 70 (1): 8–12.

Schor, Juliet, and Andrew Jorgenson. 2019. "Is it Too Late for Growth?" *Review of Radical Political Economics* 51 (2): 320–29.

Spash, Clive. 2020. "The Capitalist Passive Environmental Revolution." *Ecological Citizen* 4 (1): 63–71.

Srivastav, Sugandha, and Ryan Rafaty. 2021. *Five Worlds of Political Strategy in the Climate Movement*. Institute for New Economic Thinking at the Oxford Martin School.

Steffen, Will, Johan Rockström, Katherine Richardson, Timothy M. Lenton, Carl Folke, Diana Liverman, Colin P. Summerhayes, et al. 2018. "Trajectories of the Earth System in the Anthropocene." *Proceedings of the National Academy of Sciences* 115 (33): 8252–59.

Stephan, Maria J., and Erica Chenoweth. 2011. *Why Civil Resistance Works: The Strategic Logic of Nonviolent Conflict*. Columbia University Press.

Stern, Nicholas. 2006. *Stern Review on the Economics of Climate Change*. Cambridge University Press.

Stoner, Alexander, and Andony Melathopoulos. 2015. *Freedom in the Anthropocene: Twentieth-Century Helplessness in the Face of Climate Change*. Palgrave Macmillan.

Stuart, Diana, Gunderson, Ryan, and Brian Petersen. 2020. *Climate Change Solutions: Beyond the Capital-Climate Contradiction*. University of Michigan Press.

Stuart, Diana, Gunderson, Ryan, and Brian Petersen. 2023. *A Climate Agenda for System Change: From Theory to Social Transformation*. MayFly Press.

Taylor, Matthew. 2021. "Environment Protest Being Criminalised around World, Say Experts." *Guardian*, April 29. https://www.theguardian.com/environment/2021/apr/19/environment-protest-being-criminalised-around-world-say-experts.

Thombs, Ryan P. 2017. "The Paradoxical Relationship between Renewable Energy and Economic Growth: A Cross-National Panel Study, 1990–2013." *Journal of World-Systems Research* 23 (2): 540–64.

Thompson, Allen. 2010. "Radical Hope for Living Well in a Warmer World." *Journal of Agricultural and Environmental Ethics* 23:43–59.

Treanor, Brian. 2010. "Environmentalism and Public Virtue." *Journal of Agricultural and Environmental ethics* 23:9–28.

Wallace-Wells, David. 2018. *The Uninhabitable Earth: A Story of the Future*. Penguin.

Wiedmann, Thomas, Manfred Lenzen, Lorenz T. Keyßer, and Julia K. Steinberger. 2020. "Scientists' Warning on Affluence." *Nature Communications* 11 (1): 3107.

Williamson, Katie, Aven Satre-Meloy, Katie Velasco, and Kevin Green. 2018. *Climate Change Needs Behavior Change Making the Case for Behavioral Solutions to Reduce Global Warming: Making the Case for Behavioral Solutions to Reduce Global Warming*. Rare.

Wright, Erik Olin. 2010. *Envisioning Real Utopias*. Verso.

Wright, Erik Olin. 2019. *How to Be an Anticapitalist in the Twenty-First Century*. Verso.

York, Richard. 2016. "Decarbonizing the Energy Supply May Increase Energy Demand." *Sociology of Development* 2 (3): 265–72.

York, Richard, and Shannon Elizabeth Bell. 2019. "Energy Transitions or Additions? Why a Transition from Fossil Fuels Requires More than the Growth of Renewable Energy." *Energy Research & Social Science* 51:40–43.

Glossary

Action-oriented science: A methodological approach for doing collaborative research with practitioners and community partners that can inform practice, programs, community development, and policy while contributing to the scientific knowledge base. (Small, Stephen A., and Lynet Uttal. 2005. "Action-Oriented Research: Strategies for Engaged Scholarship." *Journal of Marriage and Family* 67 [4]: 936–48.)

Agroecology: A multifaceted approach that functions as a science, a farming method, and a movement. As a science and method, it integrates ecological principles into sustainable farming system design, promoting farm diversification and minimizing external input dependence. As a movement, it represents a radical call for social, environmental, and climate justice, particularly advocated by peasant movements like La Vía Campesina.

Alterlife: Michelle Murphy defines this term as an acknowledgment of "extensive chemical relations" —that is, between the human and more-than-human worlds—as well as a recognition of the "resurgent life" made possible via those relations. Alterlife is both the aftermath of "hurtful and deadly entanglements" and the "potential to become something else, to defend and persist, and . . . to become alter-wise." (Murphy, Michelle. 2017. "Alterlife and Decolonial Chemical Relations." *Cultural Anthropology* 32 [4]: 500.)

Animalizing: The act of becoming or embodying of nonhuman animals. In the humanities, this equates to conceptualizing how an animal thinks, feels, moves, and engages the world around them. In the social sciences, and in environmental sociology particularly, animalizing is not about the physical embodiment of an animal but the expansion of the subdiscipline to center nonhumans as those who create and share environments.

Anti-carceral(ity): A movement that decenters policing and incarceration as solutions or correctives to human behaviors that have been coded as criminal.

Capitalist realism: The contemporary, widespread inability to see that there are viable political economic alternatives to capitalism(s).

Care and repair: A movement related to degrowth through which people, objects, and the more-than-human lifeworlds are rendered as not disposable. An ethos of mending and sustaining.

City/cities: Sites of concentrated people, capital, and industry. While traditionally thought of as separate from nature, cities are socio-natural configurations in both their reliance on natural resources and the composition of their built environments.

Climate change countermovement: Fossil fuel interests and their tactics and alliances to purposefully delay and undermine climate mitigation policy.

Climate debt: A recognition that Global North nations are historically responsible for disproportionately larger contributions to climate change than Global South nations and should, therefore, compensate Global South nations for climate change-related damages and aid them with adaptation efforts.

Colonialism: A strategy of political rule that involves direct, violent, and ideological subjugation of subordinated populations and lands.

Coloniality: A term used to refer to the colonial conditions (ontological and epistemological) that shape the modern world and coincide with liberal democratic modes of state governance.

Commodity: In political economy, an object or an activity produced by human labor specifically for sale on the market. In a capitalist economy, those who own or control the means of production are structurally driven to produce commodities, from the sale of which they obtain profit.

Community-based participatory research: Research that involves meaningful and long-term collaboration between participants and researchers in the planning, conduct, application, and translation of research; more common in environmental health sciences and interdisciplinary research involving health, environmental, and social justice issues.

Complexity: A characterization of a system, model, or other process in which multiple parts change through their interaction with each other, as well as through their interaction with external influences. As parts interact over time, some become something new, others disappear, some remain stable, while still others emerge. Moreover, as parts change and acquire new properties, they often impart new properties to the entire system, model, or other process at issue, which in turn are reflected back into the parts, and so on.

Corporate food regime: A theoretical concept developed by Harriet Friedmann and Philip McMichael to analyze the current dominant form of global food provisioning (1980s onward). Following earlier British-centered (1870s–1930s) and US-centered (1950s–70s) regimes, the corporate food regime represents the third major phase in the historical development of global food systems. It is characterized by transnational corporate control, the financialization and digitalization of agriculture, and the treatment of food primarily as a commodity for profit rather than sustenance. This system is exemplified by large-scale monocultures like soybean production that prioritize export markets over local food needs.

Critical animal studies: An interdisciplinary field that interrogates the social, political, and ethical dimensions of human-nonhuman animal relationships, emphasizing the intersections of speciesism with other systems of oppression. Rooted in activist scholarship, it seeks not only to critique exploitative structures but to advance liberatory practices for all beings.

Critical public sociology: Sociological research explicitly directed toward understanding and furthering social justice goals through engagement with various publics.

De-naturalization: The practice of removing a person, group, or other entity from the category of nature. This can be done critically (e.g., by using the concept of queer kin to denaturalize heterosexual nuclear families) or violently (e.g., by removing queer relations from the range of human behavior).

Degrowth: Both a concept and a movement, degrowth describes planned reductions in energy and resources as well as the idea for creating a society that prioritizes health and well-being over economic growth and profit. Degrowth proponents seek to redirect society away from the fetishism of (economic) growth and toward socioeconomic arrangements based in sharing, maintenance, and future-oriented preservation.

Disabled ecologies: Sunaura Taylor defines this term as "the profound alterations to the capacities and functioning of an entity or system, which limits its ability to sustain itself and others as it previously had, and which alters its reproductive capacities." The term recognizes that "nature has been altered and damaged in profound and serious ways [that] resemble and are related to the disablement of human beings." (Taylor, Sunaura. 2021. "Age of Disability: On Living Well with Impaired Landscapes." *Orion Magazine*, November 10. https://orion-magazine.org/article/age-of-disability/.)

Diversity: A representation of difference in all its forms.

Doomerism: A trending belief fueled by defeatism and fatalism that it is too late to act to mitigate climate change and that apocalyptic climate impacts are inevitable.

Ecological modernization: A mainstream theory in environmental sociology that proposes addressing environmental problems through technological innovation and market-based solutions. Proponents advocate the position that economic growth can be decoupled from resource use, and that institutions and societies can modernize sustainably to allow economic growth to continue without environmental harms. The term can also refer to a program of environmental policy or a literature that analyzes environmental policies.

Ecosexuality: A practice and field of research that explores the intersections between sexology and ecology. Ecosexuals may focus on environmentally friendly sex toys or may seek out the natural world for erotic or sensual enjoyment. As an environmental activist strategy, ecosexuality stresses that humans' relationships with the more-than-human world must be non-extractive and can be a source of deep nourishment and joy.

Emancipation: Involves setting someone free from legal, social, or political restrictions. Liberation is the physical manifestation of emancipation.

Empire: A hierarchical social formation in which a state exerts direct or indirect political influence and control over subordinated territories and peoples.

Engaged public sociology: Sociological research that involves deliberate, deep, long-term engagement with impacted publics throughout the research process, prioritizing the questions and goals of impacted publics and working to develop knowledge that is useful and oriented toward improving social and environmental justice.

Environment: While the precise meaning of the word often varies, environment always has to do with an object's surroundings. These surroundings may be technological, social, political, or of some other contextual sort, though in this volume as well as in most environmental sociological conversations the word "environment" refers to the totality or some subset of the natural world. Human beings are often included, but do not have to be.

Environmental degradation: The social act of rendering an ecosystem, a specific set of ecosystems, or broader biophysical systems dramatically less amenable to complex life.

Environmental health: Refers to the ways that human health is affected by the environment (for instance, asthma from air pollution, exposure to hazardous chemicals at work, and so on).

Environmental injustice: When marginalized groups are disproportionately exposed to environmental harms that impacts their health and safety.

Environmental justice: The idea that all people deserve to have safe, healthy environments and the right to meaningfully participate in decisions that affect their well-being.

Environmental sociology: An academic field that uses sociological and ecological insights to analyze how human processes, in the widest conceivable conception of the expression, and nonhuman processes, in the widest conceivable conception of the expression, interact and change each other, in the widest conceivable conception of the expression.

Environmental violence: The Women's Earth Alliance and Native Youth Sexual Health Network define this term as "the disproportionate and often devastating impacts that the conscious and deliberate proliferation of environmental toxins and industrial development . . . have on Indigenous women, children and future generations, without regard from States or corporations for their severe and ongoing harm." (Women's Earth Alliance/Native Youth Sexual Health Network. 2016. *Violence on the Land, Violence on Our Bodies: Building an Indigenous Response to Environmental Violence*. http://landbodydefense.org/.)

Essentialism: A way of linking all members of a certain group by imposing a set of traits that are deemed as natural onto them—that is, an "essence."

Externality: A negative externality is a consequence entailed by the production of a commodity whose cost is not paid by those who derive profit from the sale of that commodity. Examples of environmental externalities include air pollution, which entails increased healthcare costs, the cost of providing water to communities whose sources have been spoiled by industrial activity, the cost of restoring damaged ecosystems, and so on. A positive externality is a benefits derived from commodity production by uninvolved third parties.

Feminism: A practice of investigating and disrupting hierarchical relations of power informed by, but not limited to, analyses of inequitable gender relations.

Food sovereignty: The right of peoples to control their own food and agricultural systems, emphasizing local production through sustainable methods. Unlike food security, which focuses on access to food, food sovereignty explicitly demands social control over the entire food system, prioritizing the needs of those who produce, distribute, and consume food over corporate market demands.

Green growth: A strategy to mitigate environmental problems using technology while continuing to prioritize economic growth (in terms of gross domestic product) as a societal goal.

Gross Domestic Product (GDP): Used as the primary indicator of the performance of a nation's economy. It is a measure of the total value of all goods and services produced within a country over a specific time period.

Imperial/colonial episteme: Refers to the imperial or colonial order of knowledge, as well as the beliefs, values, and assumptions that underpin what is considered legitimate knowledge.

Imperialism: A strategy of political rule over foreign peoples and lands that does not always involve direct conquest.

Implementation science: The scientific study of methods to promote the systematic uptake of research findings and other evidence-based practices into routine practice. (Bauer, Mark S., Laura Damschroder,

Hildi Hagedorn, Jeffrey Smith, and Amy M. Kilbourne. 2015. "An Introduction to Implementation Science for the Non-Specialist." *BMC Psychology* 3:32.)

Indigenous knowledges: Refers to culturally specific Indigenous ways of knowing that are place-based, transmitted intergenerationally, and dependent on Indigenous peoples' ongoing, active relationships with the land, more-than-human-kin, and each other. Some examples might include the transmission of knowledge through language, stories, ceremonial practices and mentorship relationships between elders and youth.

Indigenous peoples: The descendants of the original inhabitants in a particular geographic area.

Indigenous sovereignty: Can be conceptualized as the political nationhood of Indigenous peoples. In the United States and Canada, this refers to the legally mandated trust responsibility and government-to-government relationship between the federal government and Indigenous nations. Beyond this state-centric definition sovereignty also describes Indigenous people's right to exercise their cultural practices and lifeways free from the violent constraints and assimilating forces.

Interdisciplinary science: Integrates scientific knowledge across disciplines. (Benard, Marianne, and Tjard De Cock-Buning. 2014. "Moving from Monodisciplinarity towards Transdisciplinarity: Insights into the Barriers and Facilitators that Scientists Faced." *Science and Public Policy* 41 [6]: 720–33.)

Intersectionality: A concept, originating in Black feminist thought, that describes the concomitant and interlocking nature of socio-political oppression. In an intersectional analysis, a queer Black woman's minoritized social location is related to the *simultaneous*—rather than separately additive—devaluation of her racial identity, her sexuality, and her gender identity.

Legal personhood: Incorporation of a group of individuals provides legal personhood to the entity created. The corporation is thus legally separated from the individuals who founded it and who run it, and the contracts it enters into and the debt it acquires do not bind these people's individual wealth.

Liberation: The act of setting someone (human or nonhuman) free from imprisonment, confinement, or slavery.

Non-reformist reforms: Policies that change key systemic relationships and open the door for further social transformation.

Philosophical pluralism: The belief that multiple, if not necessarily all, perspectives are desirable.

Political economy: Macro-level spaces and structures where the state, businesses, labor, and civil society struggle over the control of natural and other resources.

Polysemic scholarly engagement (PSE): Goes beyond a discussion of value-free research and limited engagement in activism, to the recognition that (1) our engagement with environmental issues (and animals broadly) is likely driven by a desire to support or advocate for these communities; (2) social science is inherently human-centered, (3) nonhuman animals occupy multiple realities of human existence (companion, food, entertainment, sentinel, etc.); (4) the inclusion or exclusion of nonhuman animals has impacts on the material realities of nonhuman animal lives; (5) nonhuman animals are distinct from other sentient participants and subjects because they have few legal protections; and (6), that the inclusion of nonhuman animals is important because we share mutually experienced environmental outcomes.

Public sociology: Sociological research explicitly directed towards engagement with various publics, including public-facing translational work bringing sociology to more mainstream audiences (a.k.a. traditional public sociology) and close engagement and durable collaboration with communities or groups of stakeholders (a.k.a. organic public sociology). (Burawoy, Michael. 2005. "For Public Sociology." *American Sociological Review* 70 [1]: 4–28.)

Queering: An idea originating in queer theory, one that was once centered on gender and sexuality but has now been expanded to consider how heteronormativity and systems of power influence systems of oppression that are situated in binary thinking. Queering rejects simplistic binaries and boundaries in favor of expansive categories, connections, and challenges to oppressive systems. Relatedly, scholars have asserted that environmentalism and concern for environmental issues have largely operated on a heteronormative (hetero-ecological) platform.

Race: A social construction to describe a group of people who share physical and cultural traits as well as a common ancestry. (Golash-Boza, Tanya. 2016. "A Critical and Comprehensive Sociological Theory of Race and Racism." *Sociology of Race and Ethnicity* 2 [2]: 129–41.)

Racial capitalism: A process and structure whereby the uneven accumulation of capital occurs through racialized class stratification, ecosystem destruction, economic exploitation, and a state that supports such social, political, and environmental arrangements.

Racialization: The extension of racial meaning to a previously racially unclassified relationship, social practice, or group. (Omi, Michael, and Howard Winant. 2014. *Racial Formation in the United States*. Routledge.)

Racism: The state-sanctioned or extralegal production and exploitation of group-differentiated vulnerability to premature death. (Gilmore, Ruth Wilson. 2007. *Golden Gulag*. University of California Press.)

Reflexive research ethics: Ethical guidelines and decision-making principles that require continual reflection on research practices, potential risks and benefits, and relationships between researchers and participants;

this is particularly important when research topics involve significant uncertainty regarding developing technologies, topics, and/or methods.

Relativism: A perspective that claims all value judgments, including those that have to do with reality, knowledge, morality, and aesthetics, are only valid from the cultural perspective that holds them.

Reproductive justice: An ethos, based in Black feminist thought, that all people—whatever their medical needs—have the right to bodily autonomy, the right to parent or not parent, and the right to live and make family in safe and sustainable environments.

Repronormativity: The normalization of human reproduction and the denaturalization of pregnancy termination, contraception, or childlessness.

Risk society: Describes how modern industrial society transformed to a new era of technological hazards—where technological hazards such as pollution are distributed and affect people unequally, and risk has become a key feature of society.

Scientism: A perspective that claims social processes can be best studied by importing methods from the physical sciences.

Settler colonialism: An enduring social structure through which settlers attempt to remove Indigenous peoples and communities through legal, political, and cultural mechanisms in order to assert total control over Indigenous lands. These may include but are not limited to: physical violence, cultural assimilation, political disenfranchisement, and genocide.

Social metabolism: Similar to the metabolism of an organism, the exchanges between a society and its environment that allows it to keep its internal structure, including the gathering of materials from the earth's crust and what lives on it, that are then transformed into the objects and energy people need or want to live their lives, and the rejection of waste materials into the environment.

Sociology: The study of social processes. What is meant by the word "study" here is focused attention on developing relationships between theory, or abstract thinking, and method, or practices and procedures for inquiry, so that they can best explain the social processes of concern. What is meant by the word "social" are dynamics of any sort that have to do with human beings. These dynamics are understood as temporal, since they take place over time, as spatial, since take place across space, and as multi-scalar, since they take place across hierarchical, structural, conceptual, and other scales. Finally, what is meant by the word "processes" here is that sociologists consider their objects of study to be interactive, malleable activities and structures and not unchanging, monolithic things.

Socionature: An understanding that the "social" and the "natural" are inseparable and intertwined, and take many forms—from apparently "natural" (but human-made) parks and green spaces to apparently "social" things like skyscrapers that nevertheless rely on water, concrete, and other environmental inputs. This concept has been developed by urban political ecologists (among others), and is outlined in Heynen et al. 2006. (Henyen, Nik, Maria Kaika, and Erik Swyngedouw, eds. 2006. *In the Nature of Cities.* [Routledge].)

Sufficiency: A level of production and consumption that meets human needs without continuously increasing—having enough, rather than more and more.

Technological determinism: An approach to thinking about technology that views new technologies as a primary cause of social and historical changes on sweeping scales, as well as changes in how we think and behave on individual scales.

Technological risk: Describes both the probability and magnitude of negative effects of technological hazards on the environment, human health, and safety.

Total liberation: An applied theoretical concept that emerged in the early 2000s to recognize and unite distinct nonhuman animal, human, and environmental social justice movements seeking to address shared and overlapping oppressive systems. It is considered a commitment, rather than a blueprint.

Transdisciplinary science: Integrates scientific knowledge across disciplines like interdisciplinary science, but also incorporates societal perspectives (e.g., community members) into research in ways that both inform research design and motivate research to become more applied, or actionable. (Benard and De Cock-Buning 2014.)

Treadmill of production: A theory stating that the political economy is oriented around defending the interests of large polluting firms whose economic growth depends on the ever-increasing extraction and contamination of natural resources.

Urban imperialism: A term describing the power cities exert over their hinterlands, in terms of marshalling rural labor and resources for urban consumption. (Brechin, Gray. 2006. *Imperial San Francisco.* University of California Press.)

Urbanization: A process of the production of the built environment that involves both city-building *and* the transformation of hinterlands (e.g., through enclosures, infrastructure construction, extraction, waste disposal, etc.) in support of urban agglomerations. (Brenner, Neil. 2014. *Implosions/Explosions: Towards a Study of Planetary Urbanization.* Jovis Verlag.)

Notes on Contributors

HILLARY ANGELO is associate professor of sociology and director of the Center for Critical Urban and Environmental Studies at the University of California Santa Cruz. Her work combines historical sociology, critical social theory, and urban political economy and ecology to analyze contemporary urban and environmental culture and politics. Her first book, *How Green Became Good: Urbanized Nature and the Making of Cities and Citizens*, was published in 2021 by the University of Chicago Press. She is currently writing a book on the American West's 610 million acres of public lands, which are key sites of climate transitions and flashpoints of twenty-first-century political conflict.

J. M. BACON is an assistant professor of sociology at Grinnell College. Her work is primarily informed by the activism, art, stories, and theorizing of Indigenous peoples, environmental sociology, and scholarship on social movements.

SHANNON ELIZABETH BELL is professor of sociology at Virginia Tech. Her research focuses on just energy transitions, the socio-ecological impacts of fossil fuel extraction and transport, and the traditions, lifeways, and livelihoods that bind a diversity of Appalachian peoples to the medicinal herbs and wild foods of the Central Appalachian forests. Her work is informed by a number of literatures, including ecofeminisms, environmental justice, environmental sociology, and social and political ecology. She is author of two award-winning books: *Our Roots Run Deep as Ironweed: Appalachian Women and the Fight for Environmental Justice* (University of Illinois Press) and *Fighting King Coal: The Challenges to Micromobilization in Central Appalachia* (MIT Press).

HOLLY JEAN BUCK is an associate professor of environment and sustainability at the University at Buffalo. She is an environmental sociologist whose research focuses on public engagement with emerging climate and energy technologies, drawing from political economy, political ecology, science and technology studies, and the sociology of the future. She is the author of *After Geoengineering*

(Verso, 2019) and *Ending Fossil Fuels: Why Net Zero Is Not Enough* (Verso, 2021). She holds a PhD in development sociology from Cornell University and an MS in human ecology from Lund University.

IAN CARRILLO is an assistant professor of sociology and affiliate faculty in the Center for Brazil Studies at the University of Oklahoma. His research examines the relationship between race, class, and the environment, with a focus on Latin America and the United States. As a Mexican American, his family background instilled an interest in understanding how colonial legacies shape present-day social inequalities, including the character and structure of racial capitalism. His forthcoming book *The Business of Racism* (Duke University Press) uses the lens of racial capitalism to analyze the struggle to improve labor and environmental practices in Brazil's agribusiness sector.

ALISSA CORDNER is associate professor of sociology at Whitman College. Her research focuses on environmental sociology, environmental health and justice, the sociology of risk and disasters, and public engagement in science and policymaking. Her work draws on critical political economy theories of environmental risk, interdisciplinary research on the social construction of knowledge and ignorance, and environmental justice scholarship. She seeks to develop research that is useful to impacted communities and supports their environmental health and justice goals. She is the author of *Toxic Safety: Flame Retardants, Chemical Controversies, and Environmental Health* (Columbia University Press), and the coauthor of *The Civic Imagination: Making a Difference in American Political Life* (Paradigm).

JORDAN FOX is associate professor of sociology and of environment and sustainability at the State University of New York (SUNY) at Buffalo. His broad interest is in our limited ability to control nature, an interest he approaches through combining post-positivist historical sociology, environmental sociology, political ecology, realist philosophies, histories of science, and other conversations as needed. His forthcoming book from NYU Press, provisionally titled *Nature's Complex* and coauthored with Daniel Shtob, details how institutional ignorances of unruly ecological complexities accrues through time to produce many of our contemporary environmental problems.

PATRICK TRENT GREINER is assistant professor of sociology, eScience Institute Data Science fellow, and faculty affiliate with the Center for Studies in Demography and Ecology and the Center for Statistics in the Social Sciences at the University of Washington, Seattle. Patrick is broadly interested in developing theory and method that allow for better understandings of social-ecological interactions, and how such understandings can be made actionable for communities and policy actors. He uses quantitative, historical, and computational methods to characterize uncertainty about the complex relationships between social and natural systems, and how it is that social decision-making can develop path-dependent dynamics between social and natural processes that take focused and intentional action to change. His most central research efforts include characterizing how social inequalities pattern atmospheric pollution dynamics in built environments, clarifying challenges for energy use and grid decarbonization efforts both within the United States and internationally, identifying structural holes and knowledge gaps in the study of social-ecological interactions, and theorizing the process by which social, ecological, and technical complexes develop, stabilize, and become entrenched over time.

MATTHEW HOUSER is a senior social scientist at The Nature Conservancy's Chesapeake Bay Agriculture Program and Assistant Research Professor at University of Maryland Center for Environmental Science, Horn Point Laboratory. Drawing on political ecology/economy, environmental sociology, and transdisciplinary perspectives, Matt's research examines the cross-scale factors that shape societal recognition of and response to environmental change. He is particularly passionate about working with farmers and striving to build stronger relationships between the conservation and agricultural communities. Matt, along with his (awesome!) collaborators, have received over thirty million dollars in funding for their applied-research efforts from sources including the National Science Foundation and the United States Department of Agriculture.

CHRISTINE LABUSKI is associate professor of science, technology and society and women's, gender, and sexuality studies at Virginia Tech; she also serves as the associate director of the interdisciplinary ASPECT program. A former nurse practitioner with a PhD in cultural anthropology, she comes to environmental politics via (eco)feminist and other material approaches to justice, both of which she engages via trans- and interdisciplinary modes of analysis. She sits most comfortably at the intersections of queer and feminist STS.

AMALIA LEGUIZAMÓN is an associate professor of sociology and core faculty at the Stone Center for Latin American Studies at Tulane University. Her research and teaching integrate environmental sociology, the sociology of development, and Latin American critical social sciences to analyze resource extractivism in Latin America, environmental collective action, and the social construction of knowledge in interdisciplinary climate change research. Leguizamón's work has appeared in journals such as *The Journal of Peasant Studies, Latin American Perspectives,* and *Geoforum* and in edited volumes such as the *Routledge Handbook of Latin America and the Environment* and the *Handbook of Inequality and the Environment.* In 2017, she received the Robert Boguslaw Award for Technology and Humanism from the Environmental Sociology Section of the American Sociological Association. Her first book, *Seeds of Power: Environmental Injustice and Genetically Modified Soybeans in Argentina,* has won multiple awards, including the Best Book Prize from the Global Development Studies Section of the International Studies Association, the Allan Schnaiberg Outstanding Publication Award from the Environmental Sociology Section of the American Sociological Association, and an Honorable Mention for Best Book Award from the Environment Section of the Latin American Studies Association. *Seeds of Power* has been translated into Spanish and Portuguese.

MICHAEL WARREN MURPHY is assistant professor of Black studies and affiliated faculty in sociology at Occidental College. His research and teaching emphasize anticolonial and environmental approaches to sociological thought. More specifically, his research utilizes historical, ethnographic, spatial, and visual methods to study the socioecological implications of racialization, colonialism, and slavery in the modern world. He is also interested in the politics and sociology of racial and environmental knowledge. His first book project, tentatively titled *The Plantation Problem: Reckoning with the Environmental Significance of Race* focuses on the regime of environmental justice in the state formerly known as the Colony of Rhode Island and Providence Plantations to reveal how and why the state depends on colonial unknowing to make sense of and respond to the nexus of racial and environmental inequality in America.

J. P. SAPINSKI is associate professor of environmental studies at Université de Moncton in Canada. He is interested in how the structures of capitalism and corporate power mediate the socio-ecological metabolism, and how we can transform and decolonize this relationship to make it just and sustainable. His work is informed by diverse strands of thought, including environmental sociology, French structural anthropology, critical political economy, ecofeminism, and decolonial political ecology, among others.

DAN SHTOB is an assistant professor of sustainability and health at Michigan Technological University. His interests center on environmental change, disaster, complexity, and governance, with an applied focus both on the United States and sub-Saharan African contexts. Focusing on ways to improve disaster outcomes by better understanding complexity in both cause and effect, he employs participatory, community-engaged methods to explore both professional and personal interactions with change and its consequences. He also focuses on improving environmental and regulatory governance structures that help ensure efficient and effective use of science in environmental initiatives.

DIANA STUART is a professor in the School of Earth and Sustainability at Northern Arizona University. Coming from an academic background in the natural sciences, she became passionate about the social drivers of environmental problems and how social solutions can address these problems. Her work is now grounded in environmental sociology, with projects applying mid-century theory and, more recently, philosophy to better understand the contradictions related to global climate change. Diana's work focuses primarily on climate change drivers and solutions, but she also has projects related to conservation and biodiversity loss. She is the author of *What is Environmental Sociology?* (Polity Press) and has a book in production titled *Rejecting Climate Doomism* (University of Michigan Press).

TANESHA A. THOMAS is an assistant professor of sociology at Montclair State University in New Jersey. She has a PhD in sociology from the Graduate Center of the City University of New York. Her research and teaching are in the fields of environmental sociology, critical social theory, and public health. Growing up in an environmental justice community in Albany, New York inspired her interest in social science. She is passionate about understanding the structural forces that produce and maintain health disparities.

ABRAHAM VANSELOW graduated with a BA in sociology from Western Washington University in 2022. He now works as a paraeducator in Seattle, where he spends his time reading, biking, and figuring things out.

KIRSTEN VINYETA is an assistant professor of sociology at Utah State University. Her research foci include federal land and forest fire management, climate change, federal/Tribal relations, and multispecies justice. Over the past ten years, she has published in the areas of wildfire and forest management policy, Indigenous and settler colonial studies, and the intersectional impacts of climate change. Through much of her academic career, she has served as a collaborating settler researcher in Indigenous- and Tribe-led environmental initiatives.

CAMERON T. WHITLEY, PhD, is an associate professor in the Department of Sociology at Western Washington University in Bellingham, Washington. He studies issues concerning the environment and human-animal relationships. To date, he has over six dozen publications featured in places like *Proceedings of the National Academy of Sciences*, *Academic Emergency Medicine*, *Sociological Inquiry*, and *Annual Review of Sociology*. He received a National Science Foundation Early CAREER Award for 2023–28 to partner with zoos and aquariums across the country to assess how animal imagery can be better used to evoke emotion and engage conservation behaviors.

Index

Founded in 1893,
UNIVERSITY OF CALIFORNIA PRESS
publishes bold, progressive books and journals on topics in the arts, humanities, social sciences, and natural sciences—with a focus on social justice issues—that inspire thought and action among readers worldwide.

The UC PRESS FOUNDATION
raises funds to uphold the press's vital role as an independent, nonprofit publisher, and receives philanthropic support from a wide range of individuals and institutions—and from committed readers like you. To learn more, visit ucpress.edu/supportus.

www.ingramcontent.com/pod-product-compliance
Lightning Source LLC
Chambersburg PA
CBHW020510270326
41926CB00008B/814

* 9 7 8 0 5 2 0 4 2 1 2 5 7 *